SpringerBriefs in Electrical and Computer Engineering

For further volumes:
http://www.springer.com/series/10059

Haim Dahan • Shahar Cohen • Lior Rokach
Oded Maimon

Proactive Data Mining
with Decision Trees

Springer

Haim Dahan
Dept. of Industrial Engineering
Tel Aviv University
Ramat Aviv
Israel

Shahar Cohen
Dept. of Industrial Engineering & Management
Shenkar College of Engineering and Design
Ramat Gan
Israel

Lior Rokach
Information Systems Engineering
Ben-Gurion University
Beer-Sheva
Israel

Oded Maimon
Dept. of Industrial Engineering
Tel Aviv University
Ramat Aviv
Israel

ISSN 2191-8112   ISSN 2191-8120 (electronic)
ISBN 978-1-4939-0538-6   ISBN 978-1-4939-0539-3 (eBook)
DOI 10.1007/978-1-4939-0539-3
Springer New York Heidelberg Dordrecht London

Library of Congress Control Number: 2014931371

Printed on acid-free paper

Springer is part of Springer Science+Business Media (www.springer.com)

To our families

Preface

Data mining has emerged as a new science—the exploration, algorithmically and systematically, of data in order to extract patterns that can be used as a means of supporting organizational decision making. Data mining has evolved from machine learning and pattern recognition theories and algorithms for modeling data and extracting patterns. The underlying assumption of the inductive approach is that the trained model is applicable to future, unseen examples. Data mining can be considered as a central step in the overall knowledge discovery in databases (KDD) process.

In recent years, data mining has become extremely widespread, emerging as a discipline featured by an increasing large number of publications. Although an immense number of algorithms have been published in the literature, most of these algorithms stop short of the final objective of data mining—providing possible actions to maximize utility while reducing costs. While these algorithms are essential in moving data mining results to eventual application, they nevertheless require considerable pre- and post-process guided by experts.

The gap between what is being discussed in the academic literature and real life business applications is due to three main shortcomings in traditional data mining methods. (i) Most existing classification algorithms are 'passive' in the sense that the induced models merely predict or explain a phenomenon, rather than help users to proactively achieve their goals by intervening with the distribution of the input data. (ii) Most methods ignore relevant environmental/domain knowledge. (iii) The traditional classification methods are mainly focused on model accuracy. There are very few, if any, data mining methods that overcome all these shortcomings altogether.

In this book we present a proactive and domain-driven method to classification tasks. This novel proactive approach to data-mining, not only induces a model for predicting or explaining a phenomenon, but also utilizes specific problem/domain knowledge to suggest specific actions to achieve optimal changes in the value of the target attribute. In particular, this work suggests a specific implementation of the domain-driven proactive approach for classification trees. The proactive method is a two-phase process. In the first phase, it trains a probabilistic classifier using a supervised learning algorithm. The resulting classification model from the first-phase is a model that is predisposed to potential interventions and oriented toward maximizing a utility function the organization sets. In the second phase, it utilizes the induced classifier to suggest potential actions for maximizing utility while reducing costs.

This new approach involves intervening in the distribution of the input data, with the aim of maximizing an economic utility measure. This intervention requires the consideration of domain-knowledge that is exogenous to the typical classification task. The work is focused on decision trees and based on the idea of moving observations from one branch of the tree to another. This work introduces a novel splitting criterion for decision trees, termed maximal-utility, which maximizes the potential for enhancing profitability in the output tree.

This book presents two real case studies, one of a leading wireless operator and the other of a major security company. In these case studies, we utilized our new approach to solve the real world problems that these corporations faced. This book demonstrates that by applying the proactive approach to classification tasks, it becomes possible to solve business problems that cannot be approach through traditional, passive data mining methods.

Tel Aviv, Israel   Haim Dahan
July, 2013   Shahar Cohen
                                                                          Lior Rokach
                                                                          Oded Maimon

Contents

Chapter 1
Introduction to Proactive Data Mining

In this chapter, we provide an introduction to the aspects of the exciting field of data mining, which are relevant to this book. In particular, we focus on classification tasks and on decision trees, as an algorithmic approach for solving classification tasks.

1.1 Data Mining

Data mining is an emerging discipline that refers to a wide variety of methods for automatically, exploring, analyzing and modeling large data repositories in attempt to identify valid, novel, useful, and understandable patterns. Data mining involves the inferring of algorithms that explore the data in order to create and develop a model that provides a framework for discovering within the data previously unknown patterns for analysis and prediction.

The accessibility and abundance of data today makes data mining a matter of considerable importance and necessity. Given the recent growth of the field, it is not surprising that researchers and practitioners have at their disposal a wide variety of methods for making their way through the mass of information that modern datasets can provide.

1.2 Classification Tasks

In many cases the goal of data mining is to induce a predictive model. For example, in business applications such as direct marketing, decision makers are required to choose the action which best maximizes a utility function. Predictive models can help decision makers make the best decision.

Supervised methods attempt to discover the relationship between input attributes (sometimes called independent variables) and a target attribute (sometimes referred to as a dependent variable). The relationship that is discovered is referred to as a model. Usually models describe and explain phenomena that are hidden in the dataset and can be used for predicting the value of the target attribute based on the

H. Dahan et al., *Proactive Data Mining with Decision Trees*,
SpringerBriefs in Electrical and Computer Engineering,
DOI 10.1007/978-1-4939-0539-3_1, © The Author(s) 2014

values of the input attributes. Supervised methods can be implemented in a variety of domains such as marketing, finance and manufacturing (Maimon and Rokach 2001; Rokach 2008).

It is useful to distinguish between two main supervised models: classification (classifiers) and regression models. Regression models map the input space into a real-value domain. For instance, a regression model can predict the demand for a certain product given its characteristics. On the other hand, classifiers map the input space into pre-defined classes. Along with regression and probability estimation, classification is one of the most studied models, possibly one with the greatest practical relevance. The potential benefits of progress in classification are immense since the technique has great impact on other areas, both within data mining and in its applications. For example, classifiers can be used to classify mortgage consumers as good (full payback of mortgage on time) and bad (delayed payback).

1.3   Basic Terms

In this section, we introduce the terms that are used throughout the book.

1.3.1   Training Set

In a typical supervised learning scenario, a training set is given and the goal is to form a description that can be used to predict previously unseen examples. The training set can be described in a variety of languages. Most frequently, it is described as a bag instance of a certain bag schema. A bag instance is a collection of tuples (also known as records, rows or instances) that may contain duplicates. Each tuple is described by a vector of attribute values. The bag schema provides the description of the attributes and their domains. Attributes (sometimes called fields, variables or features) are typically one of two types: nominal (values are members of an unordered set) or numeric (values are real numbers). The instance space is the set of all possible examples based on the attributes' domain values.

The training set is a bag instance consisting of a set of tuples. It is usually assumed that the training set tuples are generated randomly and independently according to some fixed and unknown joint probability distribution.

1.3.2   Classification Task

Originally the machine learning community introduced the problem of concept learning which aims to classify an instance into one of two predefined classes. Nowadays we deal with a straightforward extension of concept learning which is known as the multi-class classification problem. In this case, we search for a function that maps the set of all possible examples into a pre-defined set of class labels which are not limited to the Boolean set. Most frequently the goal of the classifiers inducers is formally defined as follows. Given a training set with several input attributes and a nominal target attribute we can derive the goal of supervised learning which is to induce an optimal classifier with minimum generalization error. The generalization error is defined as the misclassification rate over the space distribution.

1.3.3   Induction Algorithm

An induction algorithm, sometimes referred to more concisely as an inducer (also known as a learner), is an entity that obtains a training set and forms a model that generalizes the relationship between the input attributes and the target attribute. For example, an inducer may take as input, specific training tuples with the corresponding class label, and produce a classifier.

Given the long history and recent growth of the field, it is not surprising that several mature approaches to induction are now available to the practitioner. Classifiers may be represented differently from one inducer to another. For example, C4.5 represents a model as a decision tree while Naive Bayes represents a model in the form of probabilistic summaries. Furthermore, inducers can be deterministic (as in the case of C4.5) or stochastic (as in the case of back propagation).

The classifier generated by the inducer can be used to classify an unseen tuple either by explicitly assigning it to a certain class (crisp classifier) or by providing a vector of probabilities representing the conditional probability of the given instance to belong to each class (probabilistic classifier). Inducers that can construct probabilistic classifiers are known as probabilistic inducers.

1.4   Decision Trees (Classification Trees)

Classifiers can be represented in a variety of ways such as support vector machines, decision trees, probabilistic summaries, algebraic functions, etc. In this book we focus on decision trees. Decision trees (also known as classification trees) are one of the most popular approaches for representing classifiers. Researchers from various disciplines such as statistics, machine learning, pattern recognition, and data mining have extensively studied the issue of growing a decision tree from available data.

A decision tree is a classifier expressed as a recursive partition of the instance space. The decision tree consists of nodes that form a rooted tree, meaning it is a directed tree with a node called a "root" that has no incoming edges. All other nodes have exactly one incoming edge. A node with outgoing edges is called an internal or test node. All other nodes are called leaves (also known as terminal or decision nodes). In a decision tree, each internal node splits the instance space into two or

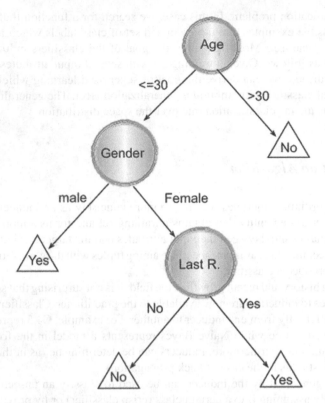

Fig. 1.1 Decision tree presenting responseto direct mailing

more sub-spaces according to a certain discrete function of the input attributes values. In the simplest and most frequent case, each test considers a single attribute, such that the instance space is partitioned according to the attribute's value. In the case of numeric attributes, the condition refers to a range.

Each leaf is assigned to one class representing the most appropriate target value. Alternatively, the leaf may hold a probability vector indicating the probability of the target attribute having a certain value. Instances are classified by navigating them from the root of the tree down to a leaf, according to the outcome of the tests along the path.

Figure 1.1 presents a decision tree that reasons whether or not a potential customer will respond to a direct mailing. Internal nodes are represented as circles while the leaves are denoted as triangles. Note that this decision tree incorporates both nominal and numeric attributes. Given this classifier, the analyst can predict the response of a potential customer (by sorting the response down the tree) to arrive at an understanding of the behavioral characteristics of the entire population of potential customers regarding direct mailing. Each node is labeled with the attribute it tests and its branches are labeled with its corresponding values.

In cases of numeric attributes, decision trees can be geometrically interpreted as a collection of hyperplanes, each orthogonal to one of the axes. Naturally, decision makers prefer less complex decision trees since they are generally considered more comprehensible. Furthermore, the tree's complexity has a crucial effect on its accuracy. The tree complexity is explicitly controlled by stopping criteria and the pruning method that are implemented. Usually the complexity of a tree is measured according to its total number of nodes and/or leaves, its depth and the number of its attributes.

Decision tree induction is closely related to rule induction. Each path from the root of a decision tree to one of its leaves can be transformed into a rule simply by conjoining the tests along the path to form the antecedent part, and taking the leaf's class prediction as the class value. For example, one of the paths in Fig. 1.1 can be transformed into the rule: "If customer age is less than or equal to or equal to 30, and the gender of the customer is 'Male'—then the customer will respond to the mail". The resulting rule set can then be simplified to improve its comprehensibility to a human user and possibly its accuracy.

Decision tree inducers are algorithms that automatically construct a decision tree from a given dataset. Typically the goal is to find the optimal decision tree by minimizing the generalization error. However, other target functions can be also defined, for instance, minimizing the number of nodes or the average depth.

Inducing an optimal decision tree from given data is considered to be a hard task. It has been shown that finding a minimal decision tree consistent with the training set is NP—hard. Moreover, it has been shown that constructing a minimal binary tree with respect to the expected number of tests required for classifying an unseen instance is NP—complete. Even finding the minimal equivalent decision tree for a given decision tree or building the optimal decision tree from decision tables is known to be NP—hard.

The above observations indicate that using optimal decision tree algorithms is feasible only for small problems. Consequently, heuristics methods are required for solving a problem. Roughly speaking, these methods can be divided into two groups, top-down and bottom-up, with clear preference in the literature to the first group.

There are various top–down decision trees inducers such as C4.5 and CART. Some consist of two conceptual phases: growing and pruning (C4.5 and CART). Other inducers perform only the growing phase.

A typical decision tree induction algorithm is greedy by nature which constructs the decision tree in a top–down, recursive manner (also known as "divide and conquer"). In each iteration, the algorithm considers the partition of the training set using the outcome of a discrete function of the input attributes. The selection of the most appropriate function is made according to some splitting measures. After the selection of an appropriate split, each node further subdivides the training set into smaller subsets, until no split gains sufficient splitting measure or a stopping criteria is satisfied.

1.5   Cost Sensitive Classification Trees

There are countless studies comparing classifier accuracy and benchmark datasets (Breiman 1996; Fayyad and Irani 1992; Buntine and Niblett 1992; Loh and Shih 1997; Provost and Fawcett 1997; Loh and Shih 1999; Lim et al. 2000). However, as Provost and Fawcett (1998) argue comparing accuracies using benchmark datasets says little, if anything, about classifier performance on real-world tasks since most research in machine learning considers all misclassification errors as having equivalent costs. It is hard to imagine a domain in which a learning system may be indifferent to whether it makes a false positive or a false negative error (Provost and Fawcett 1997). False positive (FP) and false negative (FN) are defined as follows:

$$FP = Pr\,(P|n) = \frac{Negative\ Incorrectly\ Classified}{Total\ Negative}$$

$$FN = Pr\,(N|p) = \frac{Positive\_Incorrectly\_Classified}{Total\_Positive}$$

where $\{n, p\}$ indicates the negative and positive instance classes and $\{N, P\}$ indicates the classification produced by the classifier.

Several papers have presented various approaches to learning or revising classification procedures that attempt to reduce the cost of misclassification (Pazzani et al. 1994; Domingos 1999; Fan et al. 1999; Turney 2000; Ciraco et al. 2005; Liu and Zhou 2006). The cost of misclassifying an example is a function of the predicted class and the actual class represented as a cost matrix C:

$$C(predicte\ class, actual\ class),$$

where $C(P, n)$ is the cost of false positive, and $C(N, p)$ is the cost of false negative, the misclassification cost can be calculated as:

$$Cost = FP*C(P,n) + FN*C(Np)$$

The cost matrix is an additional input to the learning procedure and can also be used to evaluate the ability of the learning program to reduce misclassification costs. While the cost can be of any type of unit, the cost matrix reflects the intuition that it is more costly to underestimate rather than overestimate how ill someone is and that it is less costly to be slightly wrong than very wrong. To reduce the cost of misclassification errors, some researchers have incorporated an average misclassification cost metric in the learning algorithm (Pazzani et al. 1994):

$$Average\ Cost = \frac{\sum_i C\,(actual\ class(i), Predicted\ Class(i))}{N}$$

Several algorithms are based on a hybrid of accuracy and classification error costs (Nunez 1991; Pazzani et al. 1994; Turney 1995; Zadrozny and Elkan 2001; Zadrozny et al. 2003) replacing the splitting criterion (i.e., information gain measurement) with a combination of accuracy and cost. For example, information cost function (ICF) selects attributes based on both their information gain and their cost (Turney 1995; Turney 2000). ICF for the i-th attribute, ICF_i, is defined as follows:

$$ICF_i = \frac{2^{\Delta I_i} - 1}{(C_i + 1)^w} \quad 0 \le w \le 1,$$

where ΔI_i is the information gain associated with the i-th attribute at a given stage in the construction of the decision tree and C_i is the cost of measuring the i-th attribute. The parameter w adjusts the strength of the bias towards lower cost attributes. When $w = 0$, cost is ignored and selection by ICF_i is equivalent to selection by ΔI_i (i.e., selection based on the information gain measure). When $w = 1$, ICF_i is strongly biased by cost.

Breiman et al. (1984) suggested the altered prior method for incorporating costs into the test selection process of a decision tree. The altered prior method, which works with any number of classes, operates by replacing the term for the prior probability, $\pi(j)$ that an example belongs to class j with an altered probability $\pi'(j)$:

$$\pi'(j) = \frac{C(j)\pi(j)}{\sum_i C(i)\pi(i)} \quad where \quad C(j) = \sum_i cost(j, i) \quad (1.1)$$

The altered prior method requires converting a cost matrix $cost(j, i)$ to cost vector $C(j)$ resulting in a single quantity to represent the importance of avoiding a particular type of error. Accurately performing this conversion is nontrivial since it depends both on the frequency of examples of each class as well as the frequency that an example of one class might be mistaken for another.

The above approaches are few of the existing main methods for dealing with cost. In general, these cost-sensitive methods can be divided into three main categories (Zadrozny et al. 2003). The first is concerned with making particular classifier learners cost-sensitive (Fan et al. 1999; Drummond and Holte 2000). The second uses Bayes risk theory to assign each example to its lowest risk class (Domingos 1999; Zadrozny and Elkan 2001; Margineantu 2002). This requires estimating class membership probabilities. In cases where costs are nondeterministic, this approach also requires estimating expected costs (Zadrozny and Elkan 2001). The third category concerns methods for converting arbitrary classification learning algorithms into cost-sensitive ones (Zadrozny et al. 2003).

Most of these cost-sensitive algorithms are focused on providing different weights to the class attribute to sway the algorithm. Essentially, however, they are still accuracy oriented. That is, they are based on a statistical test as the splitting criterion (i.e., information gain). In addition, the vast majority of these algorithms ignore any type of domain knowledge. Furthermore, all these algorithms are 'passive' in the sense that the models they extract merely predict or explain a phenomenon, rather than help users to proactively achieve their goals by intervening with the distribution of the input data.

1.6   Classification Trees Limitations

Although decision trees represent a very promising and popular approach for mining data, it is important to note that this method also has its limitations. The limitations can be divided into two categories: (a) algorithmic problems that complicate the algorithm's goal of finding a small tree and (b) problems inherent to the tree representation (Friedman et al. 1996).

Top-down decision-tree induction algorithms implement a greedy approach that attempts to find a small tree. All the common selection measures are based on one level of lookahead. Two related problems inherent to the representation structure are replication and fragmentation. The replication problem forces duplication of sub-trees in disjunctive concepts, such as $(A \cap B) \cup (C \cap D)$ (one sub-tree, either $(A \cap B)$ or $(C \cap D)$ must be duplicated in the smallest possible decision tree); the fragmentation problem causes partitioning of the data into smaller fragments. Replication always implies fragmentation, but fragmentation may happen without any replication if many features need to be tested.

This puts decision trees at a disadvantage for tasks with many relevant features. More important, when the datasets contain large number of features, the induced classification tree may be too large, making it hard to read and difficult to understand and use. On the other hand, in many cases the induced decision trees contain a small subset of the features provided in the dataset. It is important to note that the second phase of the novel proactive and domain-driven method presented in this book, considers the cost of all features presented in the dataset (including those that were not chosen for the construction of the decision tree) to find the optimal changes.

1.7   Active Learning

When marketing a service or a product, firms increasingly use predictive models to estimate the customer interest in their offer. A predictive model estimates the response probability of the potential customers in question and helps the decision maker assess the profitability of the various customers. Predictive models assist in formulating a target marketing strategy: offering the right product to the right customer at the right time using the proper distribution channel. The firm can subsequently approach those customers estimated to be the most interested in the company's product and propose a marketing offer. A customer that accepts the offer and conducts a purchase increases the firms' profits. This strategy is more efficient than a mass marketing strategy, in which a firm offers a product to all known potential customers, usually resulting in low positive response rates. For example, a mail marketing response rate of 2 % or a phone marketing response of 10 % are considered good.

Predictive models can be built using data mining methods. These methods are applied to detect useful patterns in the information available about the customers purchasing behaviors (Zahavi and Levin 1997; Buchner and Mulvenna 1998; Ling and Li 1998; Viaene et al. 2001; Yinghui 2004; Domingos 2005). Data for the models is available, as firms typically maintain databases that contain massive amounts of information about their existing and potential customers such as the customer's demographic characteristics and past purchase history.

Active learning (Cohn et al. 1994) refers to data mining policies which actively select unlabeled instances for labeling. Active learning has been previously used for facilitating direct marketing campaigns (Saar-Tsechansky and Provost 2007). In such campaigns there is an exploration phase in which several potential customers are approached with a marketing offer. Based on their response, the learner actively selects the next customers to be approached and so forth. Exploration does not come without a cost. Direct costs might involve hiring special personnel for calling customers and gathering their characteristics and responses to the campaign. Indirect costs may be incurred from contacting potential customers who would normally not be approached due to their low buying power or low interest in the product or service offer.

A well-known concept aspect of marketing campaigns is the exploration/ exploitation trade-off (Kyriakopoulos and Moorman 2004). Exploration strategies are directed towards customers as a means of exploring their behavior; exploitation strategies operate on a firm's existing marketing model. In the exploration phase, a concentrated effort is made to build an accurate model. In this phase, the firm will try, for example, to acquire any available information which characterizes the customer. During this phase, the results are analysed in depth and the best modus operandi is chosen. In the exploitation phase the firm simply applies the induced model— with no intention of improving the model—to classify new potential customers and identify the best ones. Thus, the model evolves during the exploration phase and is fixed during the exploitation phase. Given the tension between these two objectives, research has indicated that firms first explore customer behavior and then follow with an exploitation strategy (Rothaermel and Deeds 2004; Clarke 2006). The result of the exploration phase is a marketing model that is then used in the exploitation phase.

Let consider the following challenge. Which potential customers should a firm approach with a new product offer in order to maximize its net profit? Specifically, our objective is not only to minimize the net acquisition cost during the exploration phase, but also to maximize the net profit obtained during the exploitation phase. Our problem formulation takes into consideration the direct cost of offering a product to the customer, the utility associated with the customer's response, and the alternative utility of inaction. This is a binary discrete choice problem, where the customer's response is binary, such as the acceptance or rejection of a marketing offer. Discrete choice tasks may involve several specific problems, such as unbalanced class distribution. Typically, most customers considered for the exploration phase reject the offer, leading to a low positive response rate. However, an overly-simple classifier may predict that all customers in questions will reject the offer.

It should be noted that the predictive accuracy of a classifier alone is insufficient as an evaluation criterion. One reason is that different classification errors must be dealt with differently: mistaking acceptance for rejection is particularly undesirable. Moreover, predictive accuracy alone does not provide enough flexibility when selecting a target for a marketing offer or when choosing how an offer should be promoted.

For example, the marketing personnel may want to approach 30 % of the available potential customers, but the model predicts that only 6 % of them will accept the offer (Ling and Li 1998). Or they may want to personally call the first 100 most likely to accept and send a personal mailing to the next 1000 most likely to accept. In order to solve some of these problems, learning algorithms for target marketing are required not only to classify but to produce a probability estimation as well. This enables ranking the predicted customers by order of their estimated positive response probability.

Active learning merely aims to minimize the cost of acquisition, and does not consider the exploration/exploitation tradeoff. Active learning techniques do not aim to improve online exploitation. Nevertheless, occasional income is a byproduct of the acquisition process. We propose that the calculation of the acquisition cost performed in active learning algorithms should take this into consideration.

Several active learning frameworks are presented in the literature. In pool-based active learning (Lewis and Gale 1994) the learner has access to a pool of unlabeled data and can request the true class label for a certain number of instances in the pool. Other approaches focus on the expected improvement of class entropy (Roy and Mc-Callum 2001), or minimizing both labeling and misclassification costs. (Margineantu 2005). Zadrozny (2005) examined a variation in which instead of having the correct label for each training example, there is one possible label (not necessarily the correct one) and the utility associated with that label. Most active learning methods aim to reduce the generalization accuracy of the model learned from the labeled data. They assume uniform error costs and do not consider benefits that may accrue from correct classifications. They also do not consider the benefits that may be accrued from label acquisition (Turney 2000).

Rather than trying to reduce the error or the costs, Saar-Tsechansky and Provost (2007) introduced the GOAL (Goal-Oriented Active Learning) method that focuses on acquisitions that are more likely to affect decision making. GOAL acquires instances which are related to decisions for which a relatively small change in the estimation can change the preferred order of choice. In each iteration, GOAL selects a batch of instances based on their effectiveness score. The score is inversely proportional to the minimum absolute change in the probability estimation that would result in a decision different from the decision implied by the current estimation. Instead of selecting the instances with the highest scores, GOAL uses a sampling distribution in which the selection probability of a certain instance is proportional to its score.

1.8 Actionable Data Mining

There are two major issues in data mining research and applications: patterns and interest. The pattern discovering techniques include classification, association rules, outliers and clustering. Interest refers to patterns in business applications as being useful or meaningful. (Zengyou et el. 2003). One of the main reasons why we want to discover patterns in business applications is that we may want to act on them to our advantage. Patterns that satisfy this criterion of interestingness are called actionable (Silberschatz and Tuzhilin 1995; Silberschatz and Tuzhilin 1996).

Extensive research in data mining has been done on techniques for discovering patterns from the underlying data. However, most of these methods stop short of the final objective of data mining: providing possible actions to maximize profits while reducing costs (Zengyou et al. 2003). While these techniques are essential to move the data mining results to an eventual application, they nevertheless require a great deal of expert manual processing to post-process the mined patterns. Most post-processing techniques have been limited to producing visualization results, but they do not directly suggest actions that would lead to an increase of the objective utility function such as profits (Zengyou et al. 2003). Therefore it is not surprising that actionable data mining was highlighted by the Association for Computing Machinery's Special Interest Group on Knowledge Discovery and Data Mining (SIGKDD) 2002 and 2003 as one of the grand challenges for current and future data mining (Ankerst 2002; Fayyad et al. 2003).

This challenge partly results from the scenario that current data mining is a data-driven trial-and- error process (Ankerst 2002) where data mining algorithms extract patterns from converted data via some predefined models based on an expert's hypothesis. Data mining is presumed to be an automated process producing automatic algorithms and tools without human involvement and the capability to adapt to external environment constraints. However, data mining in the real world is highly constraint-based (Boulicaut and Jeudy 2005; Cao and Zhang 2006). Constraints involve technical, economic and social aspects. Real world business problems and requirements are often tightly embedded in domain-specific business rules and process. Actionable business patterns are often hidden in large quantities of data with complex structures, dynamics and source distribution. Data mining algorithms and tools generally only focus on the discovery of patterns satisfying expected technical significance. That is why mined patterns are often not business actionable even though they may be interesting to researchers. In short, serious efforts should be made to develop workable methodologies, techniques, and case studies to promote the research and development of data mining in real world problem solving (Cao and Zhang 2007).

The work presented in this book is a step toward bridging the gap described above. It presents a novel proactive approach to actionable data mining that takes in consideration domain constraints (in the form of cost and benefits), and tries to identify and suggest potential actions to maximize the objective utility function set by the organization.

1.9 Human Cooperated Mining

In real world data mining, the requirement for discovering actionable knowledge in constraint-based context is satisfied by interaction between humans (domain experts) and the computerized data mining system. This is achieved by integrating human qualitative intelligence with computational capability. Therefore, real world data mining can be presented as an interactive human-machine cooperative knowledge discovery process (known also as active/interactive information systems). With such an approach, the role of humans can be embodied in the full data mining process: from business and data understanding to refinement and interpretation of algorithms and resulting outcomes. The complexity involved in discovering actionable knowledge determines to what extent humans should be involved. On the whole, human intervention significantly improves the effectiveness and efficiency of the mined actionable knowledge (Cao and Zhang 2006). Most existing active information systems view humans as essential to data mining process (Aggarwl 2002). Interaction often takes explicit forms, for instance, setting up direct interaction interfaces to fine tune parameters. Interaction interfaces themselves may also take various forms, such as visual interfaces, virtual reality techniques, multi-modal, mobile agents, etc. On the other hand, human interaction could also go through implicit mechanisms, for example accessing a knowledge base or communicating with a user assistant agent. Interaction quality relies on performance such as user-friendliness, flexibility, run-time capability and understandability.

Although many existing active data mining systems require human involvement at different steps of process, many practitioners or users do not know how to incorporate problem specific domain knowledge into the process. As a result, the knowledge that has been mined is of little relevance to the problem at hand. This is one of the main reasons that an extreme imbalance between a massive number of research publications and rare workable products/systems has emerged (Cao 2012). The method presented in this book indeed requires the involvement of humans, namely domain experts. However, our new domain-driven proactive classification method considers problem specific domain knowledge as an integral part of the data mining process. It requires a limited involvement of the domain experts: at the beginning of the process—setting the cost and benefit matrices for the different features and at the end—analyzing the system's suggested actions.

References

Aggarwl C (2002) Toward effective and interpretable data mining by visual interaction. ACM SIGKDD Explorations Newsletter 3(2):11–22

Ankerst M (2002) Report on the SIGKDD-2002 Panel—the perfect data mining tool: interactive or automated? ACM SIGKDD Explorations Newsletter 4(2):110–111

Boulicaut J, Jeudy B (2005) Constraint-based data mining, the data mining and knowledge discovery handbook, Springer, pp 399–416

Breiman L, Friedman JH, Olshen R A, Stone C J (1984). Classification and regression trees. Wadsworth & Brooks/Cole Advanced Books & Software, Monterey, CA. ISBN 978-0-412-04841-8

Breiman L (1996) Bagging predictors. Mach Learn 24:123–140

Büchner AG, Mulvenna MD (1998) Discovering internet marketing intelligence through online analytical web usage mining. ACM Sigmod Record 27(4):54–61

Buntine W, Niblett T (1992) A further comparison of splitting rules for decision-tree induction. Mach Learn 8:75–85

Cao L, Zhang C (2006) Domain-driven actionable knowledge discovery in the real world. PAKDD2006, pp 821–830, LNAI 3918

Cao L, Zhang C (2007) The evolution of KDD: towards domain-driven data mining, international. J Pattern Recognit Artif intell 21(4):677–692

Cao L (2012) Actionable knowledge discovery and delivery. Wiley Interdiscip Rev Data Min Knowl Discov 2:149–163

Ciraco M, Rogalewski M, Weiss G (2005) Improving classifier utility by altering the misclassification cost ratio. In: Proceedings of the 1st international workshop on utility-based data mining, Chicago, pp 46–52

Clarke P (2006) Christmas gift giving involvement. J Consumer Market 23(5):283–291

Cohn D, Atlas L, Ladner R (1994) Improving generalization with active learning. Mach Learn 15(2):201–221

Domingos P (1999) MetaCost: a general method for making classifiers cost sensitive. In: Proceedings of the Fifth International Conference on Knowledge Discovery and Data Mining, ACM Press, pp 155–164

Domingos P (2005) Mining social networks for viral marketing. IEEE Intell Syst 20(1):80–82

Drummond C, Holte R (2000) Exploiting the cost (in)sensitivity of decision tree splitting criteria. In Proceedings of the 17th International Conference on Machine Learning, 239–246

Fan W, Stolfo SJ, Zhang J, Chan PK (1999) AdaCost: misclassification cost-sensitive boosting. In: Proceedings of the 16th international conference machine learning, pp 99–105

Fayyad U, Irani KB (1992) The attribute selection problem in decision tree generation. In Proceedings of tenth national conference on artificial intelligence. AAAI Press, Cambridge, pp 104–110

Fayyad U, Shapiro G, Uthurusamy R (2003) Summary from the KDD-03 panel—data mining: the next 10 years. ACM SIGKDD Explor Newslett 5(2) 191–196

Friedman JH, Kohavi R, Yun Y (1996) Lazy decision trees. In: Proceedings of the national conference on artificial intelligence, pp. 717–724

Kyriakopoulos K, Moorman C (2004) Tradeoffs in marketing exploitation and exploration strategies: the overlooked role of market orientatio. Int J Res Market 21:219–240

Lewis D, Gale W (1994) A sequential algorithm for training text classifiers. In Proceedings of the international ACM-SIGIR conference on research and development in information retrieval, pp 3–12

Lim TS, Loh WY, Shih YS (2000) A comparison of prediction accuracy, complexity, and training time of thirty-three old and new classification algorithms. Mach Learn 40(3):203–228

Ling C, Li C (1998) Data mining for direct marketing: problems and solutions. In Proceedings 4th international conference on knowledge discovery in databases (KDD-98), New York, pp 73–79

Liu XY, Zhou ZH (2006) The influence of class imbalance on cost-sensitive learning: an empirical study. In Proceedings of the 6th international conference on data mining, pp. 970–974

Loh WY, Shih X (1997) Split selection methods for classification trees. Stat Sinica 7:815–840

Loh WY, Shih X (1999) Families of splitting criteria for classification trees. Stat Comput 9:309–315

Maimon O, Rokach L (2001) Data mining by attribute decomposition with semiconductor manufacturing case study. In: Braha D (ed) Data mining for design and manufacturing, pp 311–336

Margineantu D (2002) Class probability estimation and cost sensitive classification decisions. In: Proceedings of the 13th european conference on machine learning, 270–281

Margineantu D (2005) Active cost-sensitive learning. In Proceedings of the nineteenth international joint conference on artificial intelligence, IJCAI–05

Nunez M (1991) The use of background knowledge in decision tree induction. Mach Learn 6(3): 231–250

Pazzani M, Merz C, Murphy P, Ali K, Hume T, Brunk C (1994) Reducing misclassification costs. In: Proceedings 11th international conference on machine learning. Morgan Kaufmann, pp 217–225

Provost F, Fawcett T (1997) Analysis and visualization of classifier performance comparison under imprecise class and cost distribution. In: Proceedings of KDD-97. AAAI Press, pp 43–48

Provost F, Fawcett T (1998) The case against accuracy estimation for comparing induction algorithms. In: Proceedings 15th international conference on machine learning. Madison, pp 445–453

Rokach L (2008) Mining manufacturing data using genetic algorithm-based feature set decomposition. Int J Intell Syst Tech Appl 4(1):57–78

Rothaermel FT, Deeds DL (2004) Exploration and exploitation alliances in biotechnology: a system of new product development. Strateg Manage J 25(3):201–217

Roy N, McCallum A (2001) Toward optimal active learning through sampling estimation of error reduction. In Proceedings of the international conference on machine learning

Saar-Tsechansky M, Provost F (2007) Decision-centric active learning of binary-outcome models. Inform Syst Res 18(1):4–22

Silberschatz A, Tuzhilin A (1995) On subjective measures of interestingness in knowledge discovery. In Proceedings, first international conference knowledge discovery and data mining, pp 275–281

Silberschatz A, Tuzhilin A (1996) What makes patterns interesting in knowledge discovery systems, IEEE Trans. Know Data Eng 8:970–974

Turney P (1995) Cost-sensitive classification: empirical evaluation of hybrid genetic decision tree induction algorithm. J Artif Intell Res 2:369–409

Turney P (2000) Types of cost in inductive concept learning. In Proceedings of the ICML'2000 Workshop on cost sensitive learning Stanford, pp 15–21

Viaene S, Baesens B, Van Gestel T, Suykens JAK, Van den Poel D, Vanthienen J, De Moor B, Dedene G (2001) Knowledge discovery in a direct marketing case using least squares support vector machine classifiers. Int J Intell Syst 9:1023–1036

Yinghui Y (2004) New data mining and marketing approaches for customer segmentation and promotion planning on the Internet, Phd Dissertation, University of Pennsylvania, ISBN 0-496-73213-1

Zadrozny B, Elkan C (2001) Learning and making decisions when costs and probabilities are both unknown. In Proceedings of the seventh international conference on knowledge discovery and data mining (KDD'01)

Zadrozny B, Langford J, Abe N (2003) Cost-sensitive learning by cost-proportionate example weighting. In ICDM (2003), pp 435–442

Zadrozny B (2005) One-benefit learning: cost-sensitive learning with restricted cost information. In Proceedings of the workshop on utility-based data mining at the eleventh ACM SIGKDD international conference on knowledge discovery and data mining

Zahavi J, Levin N (1997) Applying neural computing to target marketing. J Direct Mark 11(1):5–22

Zengyou He, Xiaofei X, Shengchun D (2003) Data mining for actionable knowledge: A survey. Technical report, Harbin Institute of Technology, China. http://arxiv.org/abs/ cs/0501079. Accessed 13 Jan 2013.

Chapter 2
Proactive Data Mining: A General Approach and Algorithmic Framework

In the previous section we presented several important data mining concepts. In this chapter, we argue that with many state-of-the-art methods in data mining, the overly-complex responsibility of deciding on this action or that is left to the human operator. We suggest a new data mining task, proactive data mining. This approach is based on supervised learning, but focuses on actions and optimization, rather than on extracting accurate patterns. We present an algorithmic framework for tackling the new task. We begin this chapter by describing our notation.

2.1 Notations

Let $A = \{A_1, A_2, \ldots, A_k\}$ be a set of explaining attributes that were drawn from some unknown probability distribution p_0, and $D(A_i)$ be the domain of attribute A_i. That is, $D(A_i)$ is the set of all possible values that A_i can receive. In general, the explaining attributes may be continuous or discrete. When A_i is discrete, we denote by $a_{i,j}$ the j-th possible value of A_i, so that $D(A_i) = \{a_{i,1}, a_{i,2}, \ldots a_{i,|D(Ai)|}\}$, where $|D(A_i)|$ is the finite cardinality of $D(A_i)$. We denote by $D = D(A_1) \times D(A_2) \times \ldots \times D(A_k)$ the Cartesian product of $D(A_1)$, $D(A_2)$, $\ldots$, $D(A_k)$ and refer to it as the input domain of the task. Similarly, let T be the target attribute, and $D(T) = \{c_1, c_2, \ldots c_{|D(T)|}\}$ the discrete domain of T. We refer to the values in $D(T)$ as the possible classes (or results) of the task. We assume that T depends on D, usually with an addition of some random noise.

Classification is a supervised learning task, which receives training data, as input. Let $< X;Y > = < x_{1,n}, x_{2,n}, \ldots, x_{k,n}; y_n >$, for $n = 1, 2, \ldots, N$ be a training set of N classified records, where $x_{i,n} \in D(A_i)$ is the value of the i-th explaining attribute in the n-th record, and $y_n \in D(T)$ is the class relation of that record. Typically, in a classification task, we search for a model—a function $f: D \to D(T)$, so that given $x \in D$, a realization of the explaining attributes, randomly drawn from the joint, unknown probability distribution function of the explaining attributes, and $y \in D(T)$, the corresponding class relation, the probability of correct classification, $\Pr[f(x) = y]$,

H. Dahan et al., *Proactive Data Mining with Decision Trees,*
SpringerBriefs in Electrical and Computer Engineering,
DOI 10.1007/978-1-4939-0539-3_2, © The Author(s) 2014

is maximized. This criterion is closely related to the accuracy[1] of the model. Since the underlined probability distributions are unknown, the accuracy of the model is estimated by an independent dataset for testing, or through a cross-validation procedure.

2.2   From Passive to Proactive Data Mining

Data mining algorithms are used as part of the broader process of knowledge-discovery. The role of the data-mining algorithm, in this process, is to extract patterns hidden in a dataset. The extracted patterns are then evaluated and deployed. The objectives of the evaluation and deployment phases include decisions regarding the interest of the patterns and the way they should be used (Kleinberg et al. 1998; Cao 2006; Cao and Zhang 2007; Cao 2010, 2012).

While data mining algorithms, particularly those dedicated to supervised learning, extract patterns almost automatically (often with the user making only minor parameter settings), humans typically evaluate and deploy the patterns manually. In regard to the algorithms, the best practice in data mining is to focus on description and prediction and not on action. That is to say, the algorithms operate as *passive* "observers" on the underlying dataset while analyzing a phenomenon (Rokach 2009). These algorithms neither affect nor recommend ways of affecting the real world. The algorithms only report to the user on the findings. As a result, if the user chooses not to act in response to the findings, then nothing will change. The responsibility for action is in the hands of humans. This responsibility is often overly complex to be handled manually, and the data mining literature often stops short of assisting humans in meeting this responsibility.

Example 2.1 In marketing and customer relationship management (CRM), data mining is often used for predicting customer lifetime value (LTV). Customer LTV is defined as the net present value of the sum of the profits that a company will gain from a certain customer, starting from a certain point in time and continuing through the remaining lifecycle of that customer. Since the exact LTV of a customer is revealed only after the customer stops being a customer, managing existing LTVs requires some sort of prediction capability. While data mining algorithms can assist in deriving useful predictions, the CRM decisions that result from these predictions (for example, investing in customer retention or customer-service actions that will maximize her or his LTV) are left in the hands of humans.

In *proactive* data mining we seek automatic methods that will not only describe a phenomenon, but also recommend actions that affect the real world. In data mining, the world is reflected by a set of observations. In supervised learning tasks, which are the focal point of this book, each observation presents an instance of the explaining

---

[1] In other cases, rather than maximal accuracy, the objective is minimal misclassification costs or maximal lift.

attributes and the corresponding target results. In order to affect the world and to assess the impact of actions on the world, the data observations must encompass certain changes. We discuss these changes in the following section.

2.3   Changing the Input Data

In this book, we focus on supervised learning tasks, where the user seeks to generalize a function that maps explaining attribute values to target values. We consider the training record, $< x_{1,n}, x_{2,n}, \ldots, x_{k,n}; y_n >$, for some specific n. This record is based on a specific object in the real world. For example, $x_{1,n}, x_{2,n}, \ldots, x_{k,n}$ may be the explaining attributes of a client, and y_n, the target attribute, might describe a result that interests the company, whether the client has left or not.

It is obvious that some results are more beneficial to the company than others, such as a profitable client remaining with the company rather than leaving it or those clients with high LTV are more beneficial than those with low LTV. In proactive data mining, our motivation is to search for means of actions that lead to desired results (i.e., desired target values).

The underlying assumption in supervised learning is that the target attribute is a dependent variable whose values depend on those of the explaining attributes. Therefore, in order to affect the target attribute towards the desired, more beneficial, values, we need to change the explaining attributes in such a way that target attributes will receive the desired values.

Example 2.2   Consider the supervised learning scenario of churn prediction, where a company observes its database of clients and tries to predict which clients will leave and which will remain loyal. Assuming that most of the clients are profitable to the company, the motivation in this scenario is churn prevention. However, the decision of a client about whether to leave or not may depend on other considerations, such as her or his price plan. The client's price plan, hardcoded in the company's database, is often part of the churn-prediction models. Moreover, if the company seeks for ways to prevent a client from leaving, it can consider changing the price plan of the client as a churn-prevention action. Such action, if taken, might affect the value of an explaining attribute towards a desired direction.

When we refer to "changing the input data", we mean that in proactive data mining we seek to implement actions that will change the values of the explaining attributes and consequently lead to a desired target value. We do not consider any other sort of action because it is external to the domain of the supervised learning task. To look at the matter in a slightly different light, the objective in proactive data mining is *optimization*, and not prediction. In the following section we focus on the required domain knowledge that results from the shift to optimization, and we define an *attribute changing cost* function and a *benefit* function as crucial aspects of the required domain knowledge.

2.4 The Need for Domain Knowledge: Attribute Changing Cost and Benefit Functions

The shift from supervised learning to optimization requires us to consider additional knowledge about the business domain, which is exogenous to the actual training records. In general, the additional knowledge may cover various underlying business issues behind the supervised learning task, such as: What is the objective function that needs to be optimized? What changes in the explaining attributes can and cannot be achieved? At what cost? What are the success probabilities of attempts to change the explaining attributes? What are the external conditions, under which these changes are possible? The exact form of the additional knowledge may differ, depending on the exact business context of the task. Specifically, in this book we consider a certain form of additional knowledge that consists of attribute changing costs and benefit functions. Although we describe these functions below as reasonable and crucial considerations for many scenarios, nevertheless, one might have to consider additional aspects of domain knowledge, or maybe even different aspects, depending on the particular business scenario being examined.

The attribute changing cost function, $C: D \times D \to R$, assigns a real value cost for each possible change in the values of the explaining attributes. If a particular change cannot be achieved (e.g., changing the gender of a client, or making changes that conflict with laws or regulations), the associated costs are infinite. If for some reason the cost of an action depends on attributes that are not included in the set of explaining attributes, we include these attributes in D, and call them *silent attributes*—attributes that are not used by the supervised learning algorithms, but are included in the domain of the proactive data mining task.

The benefit function $B: D \times D(T) \to R$ assigns a real value benefit (or outcome) that represents the company's benefit from any possible record. The benefit from a specific record depends not only on the value of the target attribute, but also on the values of the explaining attributes. For example, benefit from a loyal client depends not only on the target value of churning $= 0$, but also on the explaining attributes of the client, such as his or her revenue. As in the case of the attribute changing cost function, the domain D may include silent attributes. In the following section we combine the benefit and the attribute changing functions and formally define the objective of the proactive data mining task.

2.5 Maximal Utility: The Objective of Proactive Data Mining Tasks

The objective in proactive data mining is to find the *optimal decision making policy*. A policy is a mapping $O: D \to D$ that defines the impact of some actions on the values of the explaining attributes. In order for a policy to be optimal, it should maximize the expected value of a utility function. The utility function that we consider in this book results from the benefit and attribute changing cost functions in the following manner: the addition to the benefit due to the move minus the attribute changing cost that is associated with that move.

It should be noted that the stated objective is to find an optimal policy. The optimal policy may depend on the probability distribution of the explaining attributes which is considered unknown. We use the training set as the empirical distribution, and search for the optimal actions with regard to that dataset. That is, we search for the policy that, if followed, will maximize the sum of the utilities that are gained from the N training observations.

It should be also noted that the cost, which is associated to O, can be calculated directly from the function C. The cost of a *move*—that is, changing the values of the explaining attributes from $x_i = < x_{1,i}, x_{2,i}, \ldots, x_{k,i} >$ to $x_j = < x_{1,j}, x_{2,j}, \ldots, x_{k,j} >$ is simply $C(x_i, x_j)$. However, in order to evaluate the benefit that is associated with the move, we must also know the impact of the change on the target attribute. This observation leads to our algorithmic framework for proactive data mining which we present in the following section.

2.6   An Algorithmic Framework for Proactive Data Mining

In order to evaluate the benefit of a move, we must know the impact of a change on the value of the target attribute. Fortunately, the problem of evaluating the impact of the values of the explaining attributes on the target attribute is well-known in data mining and is solved by supervised learning algorithms. Similarly our algorithmic framework for proactive data mining also uses a supervised learning algorithm for evaluating impact. Our framework consists of the following phases:

1. Define the explaining attributes and the target result as in the case of any supervised-learning task.
2. Define the benefit and the attribute changing cost functions.
3. Extract patterns that model the dependency of the target attribute on the explaining attributes by using a supervised learning algorithm.
4. Using the results of phase 3, optimize by finding the changes in values of the explaining attributes that maximize the utility function.

The main question regarding phase 3 is what supervised algorithm to use. One alternative is to use an existing algorithm, such as a decision-tree (which we use in the following chapter). Most of the existing supervised learning algorithms are built in order to maximize the accuracy of their output model. This desire to obtain maximum accuracy, which in the classification case often takes the form minimizing the 0–1 loss, does not necessarily serve the maximal-utility objective that we defined in the previous section.

Example 2.3 Consider a supervised learning scenario in which a decision tree is being used to solve a question of churn prediction. Let us consider two possible splits: (a) according to client gender, and (b) according to the client price plan. It might be the case (although typically this is not the case) that splitting according to client gender results in more homogeneous sub-populations of clients than splitting according to the client price plan. Although contributing to the overall accuracy of the output decision tree, splitting according to client gender provides no opportunity for evaluating the consequences of actions, since the company cannot act to change that gender. On the other hand, splitting according to the client price plan, even if inferior in terms of accuracy, allows us to evaluate the consequences of an important action: changing a price plan.

Another alternative for a supervised learning algorithm is to design an algorithm that will enable us to find better changes in the second phase, that is, to design an algorithm that is sensitive to the utility function and not to accuracy. In Chap. 3 we propose a decision tree algorithm that displays these characteristics in regard to classification scenarios. Then, in chap. 4 we demonstrate that this alternative can contribute to the accumulated utility of the overall proactive data-mining task.

2.7   Chapter Summary

We observed in this chapter that data mining in general and supervised learning tasks in particular, tends to operate in a passive way. Accordingly, we defined a new data mining task, proactive data mining. We showed that shifting from supervised learning to proactive data mining requires additional domain knowledge. We focused on two aspects of such knowledge: the benefit function and the attribute changing cost function. Based on these two functions, we formally defined the task of proactive data mining as finding the actions, which maximize utility. We defined utility as benefit minus cost. We concluded the chapter by describing an algorithmic framework for proactive data mining.

References

Cao L (2006) Domain driven actionable knowledge discovery in real world, PAKDD2006. pp 1021–1030

Cao L (2010) Domain driven data mining, challenges and prospects. IEEE Trans Knowl Data Eng 22(6):755–769

Cao L (2012) Actionable knowledge discovery and delivery. WIREs Data Min Knowl Discove 2(2):149–163

Cao L, Zhang C (2007) The evolution of KDD: towards domain-driven data mining. Int J Pattern Recognit Artif Intell 21(4):677–692

Kleinberg J, Papadimitriou C, Raghavan P (1998) A microeconomic view of data mining. Knowl Discov Data Min 2(4):311–324

Rokach L (2009) Collective-agreement-based pruning of ensembles. Comput Stat Data Anal 53(4):1015–1026

Chapter 3
Proactive Data Mining Using Decision Trees

In the previous chapter we introduced the task of proactive data mining and sketched an algorithmic framework for solving the task: first build a prediction model and then use it for optimization. In this chapter, we focus on decision tree classifiers and describe in detail two possible ways of implementing proactive data mining using: (a) a ready-made decision tree algorithm, and (b) a novel decision tree algorithm. We designed this latter algorithm to support the optimization phase of the proposed framework.

3.1 Why Decision Trees?

Decision trees are simple yet effective techniques for predicting and explaining the relationship between the explaining attributes and the target value. Simplicity is one reason that led us to choose decision trees as the principal modeling approach and test bed for the new, proactive data mining task. In addition to their simplicity, decision trees explicitly describe the functional dependencies of the target attribute on the explaining attributes. (These dependencies are represented by splits according to the values of the explaining attributes). In relation to proactive data mining, this descriptive property of decision trees has three advantages:

1. It helps us produce recommendations on action that users can easily understand (Ben-Shimon et al. 2007)
2. It allows us to automatically and systematically search a tree for advantageous moves.
3. It preserves the privacy of the clients (Kisilevich et al. 2010; Matatov et al. 2010)

Example 3.1 example demonstrates this.

Example 3.1 Consider the decision tree in Fig. 3.1, which describes the churning patterns of the clients of a telecommunications service provider. The tree describes the relations between explaining attributes (Package, Sex and Monthly Rate) and a target attribute, reflecting whether a client left the company or remained loyal. It can be seen, for example that if we act to change the monthly voice rate for male clients

H. Dahan et al., *Proactive Data Mining with Decision Trees,*
SpringerBriefs in Electrical and Computer Engineering,
DOI 10.1007/978-1-4939-0539-3_3, © The Author(s) 2014

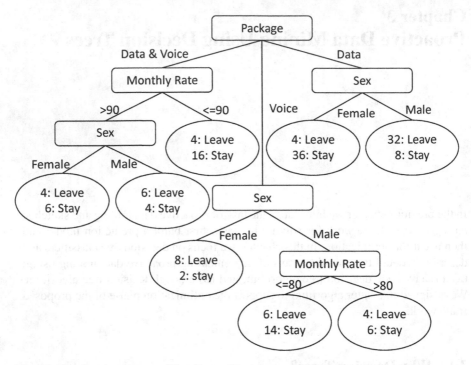

Fig. 3.1 The churning patterns of the clients of the telecommunications service provider

from a range that exceeds $ 80 to a range below $ 80, the churning probability is expected to decline. It can intuitively be seen that any two branches in the decision tree span two possible actions: moving from the first branch to the second and vice versa. Therefore, if we want to act effectively in scenarios similar to that presented in this example, we simple scan the tree for pairs of branches.

3.2   The Utility Measure of Proactive Decision Trees

Let $DT = (V,E)$ be a decision tree with the set of vertices (or nodes) $V = \{v_0, v_1, \ldots, v_{|V|}\}$; where $|V|$ is the finite cardinality of V and the set of edges (arcs) E; and where each $e \in E$ is an ordered pair of vertices: $e = <v_i, v_j>$ indicating that v_j is a direct son of v_i. We denote the decision tree's root by v_0, and assume that each vertex in V, except for v_0, has exactly one parent and either zero or more than a one direct sons (i.e., DT is indeed a tree). We consider decision trees that were trained based on the training set: $<X;Y> = <x_{1,n}, x_{2,n}, \ldots, x_{k,n}; y_n>$, for $n = 1, 2, \ldots, N$, which was drawn from the input domain under the unknown probability distribution function p_0 (and the unknown distribution of the target, given the explaining attributes).

Let $|v_i(<X;Y>)|$ be the number of records in $<X;Y>$ that reach the vertex v_i, when sorted by DT in a top-down manner. We refer to $|v_i(<X;Y>)|$ as the *size* of the vertex v_i. Let us define $p_0(c_j, v_i)$ as the estimated proportion of cases in v_i that belong to class c_j. We calculate $p_0(c_j, v_i)$ according to Laplace's law of succession:

$$p_0\left(c_j, v_i\right) = \frac{m\left(c_j, v_i\right) + 1}{|v_i\left(X;Y\right)| + 2}$$

where $m(c_j, v_i)$ is the number of records in $< X;Y >$ that reach the vertex v_i and relate to class c_j. We refer to nodes with no direct sons as leaves (or terminals) and denote the set of leaf nodes by L. We define a branch in the tree as follows.

Definition 3.1 A branch, β in the decision tree DT, is a sequence of nodes $v(0), v(1), v(2), .., v(|\beta|)$, where $|\beta|$ is the length (number of nodes) of the branch, so that:

1. $v(0) = v_0$ (i.e., $v(0)$ is the decision-tree's root)
2. For all $i = 0, 1, \ldots, |\beta|$-1, $v(i + 1)$ is a direct son of $v(i)$
3. $v(|\beta|) \in L$

Given the benefit function $B: D \times D(T) \rightarrow R$, which was defined in the previous chapter, we can define the total benefit of a branch as the sum of the benefits of all the observations that if sorted down the tree, reach the branch's terminal. We denote the total benefit of the branch β by $TB(\beta)$, and use it to assess the attractiveness of the branch. We denote the total benefit of the tree DT as:

$$TB\left(DT\right) = \sum_{\beta \in DT} TB\left(\beta\right).$$

Given a decision tree, we can examine the expected consequences of proactively acting on the records of a certain branch in order to change the values of their explaining attributes and move them to some different branch of the tree. Let β_1 and β_2 be the source and destination branches respectively (i.e., we change the values of the records in β_1, in order to move the records to β_2). The estimated merit of moving from β_1 to β_2, is the difference in the total benefits of β_2 to β_1, minus the cost that derives from the value change that is required in order to move from β_1 to β_2:

$$\text{merit}\left(\beta_1, \beta_2\right) = TB\left(\beta_2\right) \cdot \frac{|v\left(|\beta_1|\right)\left(X;Y\right)|}{|v\left(|\beta_2|\right)\left(X;Y\right)|} - TB\left(\beta_1\right) - C\left(D_1, D_2\right),$$

where D_1 and D_2 are the sub-domains of D that correspond to the terminals of β_1 and β_2, respectively and C is the attribute-changing cost function, which was defined in the previous section. The term $|v(|\beta_1|)(< X;Y >)|/|v(|\beta_2|)(< X;Y >)|$ normalizes the total benefit of β_2 to the number of records that are currently in β_1. We refer to moves from one branch to another as single-branch moves.

In principle, it is possible to exhaustively scan any given decision tree and to search for all the single-branch moves that have a positive, associated merit. However, when assessing the attractiveness of a branch, we must consider the number of records in it. A branch that has a small number of records is inherently less certain than a branch with large number of records. More specifically, even if moving from β_1 to β_2 is associated with positive merit, knowing that there are only few records in β_2 we might want to avoid the move, since we cannot be sure that the new records, coming from β_1, will behave similarly to those already in β_2. We take the number of records in a branch into consideration by adding a weight to each possible single-branch move. We denote the weight associated to the move from β_1 to β_2 as $w(\beta_1,\beta_2)$, and define the utility of a single-branch move, as follows:

$$\text{utility }(\beta_1, \beta_2) = \left[TB\,(\beta_2) \cdot \frac{|v\,(|\beta_1|)\,(X;Y)|}{|v\,(|\beta_2|)\,(X;Y)|} - TB\,(\beta_1) \right] \cdot w\,(\beta_1, \beta_2) - C\,(D_1, D_2),$$

and consider the move as advantageous, if utility(β_1,β_2) exceeds some pre-defined threshold. Notice that the weight multiplies the difference in total benefits between β_2 and β_1 and not the attribute-change cost, because the costs are considered to be (and are often almost) deterministic, whereas the total benefit depends by definition on the unknown target class distribution. In this book, we use the lower bound of a $1 - \alpha = 95\,\%$ confidence interval, on the probability of the majority class as the weight (Menahem et al. 2009; Rokach 2009). That is, denoting the majority class by c^*, we use the following weight:

$$w\,(\beta_1, \beta_2) = p_0\,(c^*, v\,(|\beta_2|)) - Z_{1-\frac{\alpha}{2}} \cdot \sqrt{\frac{p_0\,(c^*, v\,(|\beta_2|))\,(1 - p_0\,(c^*, v\,(|\beta_2|)))}{|v\,(|\beta_2|)\,(X;Y)|}}.$$

We use these weights as a heuristic, noting that the fewer the observations there are in β_2, the smaller the lower bound of the confidence interval and the more significant the suppression of the differences in the total benefits. We do not allow negative weights and in practice use the maximum between $w(\beta_1,\beta_2)$ and zero as the weight. Example 3.2 demonstrates the definitions of this section.

Example 3.2 Let us reconsider the decision tree of Example 3.1. We number the nodes of the tree in a breadth-first search (BFS) order so that the root is v0 and its sons are denoted by v1–v3 in a left-to-right manner, and so on. The decision tree in Fig. 3.1 was trained on the toy dataset of a service provider. It included 160 observations, comprising 68 clients who left and 92 who remained. The customers are described by three explaining attributes: A1—the customer's package, which can take the values: 'Data', 'Voice' and 'Data&Voice'; A2—the customer's sex, which can be either 'Female' or 'Male'; and A3—the customer's monthly rate in US dollars, with the following possible values: 75, 80, 85, 90 and 95.

It should be noted, for example, that $v_{10}(<X,Y>) = 10$ (there are 10 training observations that reach the left-most node, *Data&Voice*, which is higher than 90 monthly-rate female customers, denoted by v_{10}). Based on Laplace's law of succession, we estimate that $p_0(Leave, v_{10}) = (4 + 1)/(10 + 2)$, and $p_0(Stay, v_{10}) = (4 + 1)/(10 + 2)$.

Considering possible changes in the values of the explaining attributes, we can clearly assume that since the company cannot affect A_2, the associated cost is infinite. To illustrate, let us also assume that due to regulations the company cannot reduce the monthly rate (A_3) and the magnitude of the reduction stands for the cost of

Table 3.1 Cost matrix for changes in the customer's package

	Data	Voice	Data & Voice
Data	0	5	10
Voice	0	0	5
Data & Voice	0	0	0

the reduction (the cost of increase is infinite). The changes in A_1 are described by Table 3.1

Notice that the description above specifies an attribute changing cost function (there is a real-number corresponding cost for each possible move). Finally, let us assume that the benefit is the value of a monthly rate for a customer who remains with the company and minus that value for a customer who leaves. Notice that this specifies a benefit function.

Based on these definitions, we can focus on the branches $\beta_1 = v_0, v_2, v_7, v_{12}$ and $\beta_2 = v_0, v_2, v_7, v_{13}$. The total benefit of β_1 depends on the distribution of A_3 in this branch, since some of the clients that are sorted to β_1 have a monthly rate of 80 while others have a monthly rate of 75. Let us take the average of these values, 77.5, for demonstration purposes (in implementations, the benefit must be calculated on a per client basis). Based on the 77.5 assumption, the total benefit of β_1 is: $TB(\beta_1) = 77.5 \cdot (14 - 6) = 620$. Similarly, if we assume that the monthly rate in β_2 is 90 (the average of 85, 90 and 95), we can see that the total benefit of β_2 is $TB(\beta_2) = 90 \cdot (6 - 4) = 180$. We can easily calculate the total benefit of the entire tree by summing all the total benefits of the seven branches of the tree.

One of the possible actions that may be taken in regard to clients that are currently sorted to β_1 is to change their monthly rate, which will move them to β_2. We can also change the monthly rate of clients in β_2 in order to move them to β_1. Since the first move involves increasing the monthly rate, which has an infinite cost, clearly the corresponding merit is negative. We can compute the merit of the second move, as follows:

$$\text{merit}(\beta_2, \beta_1) = 620 \cdot \frac{10}{20} - 180 - 10 \cdot (90 - 77.5) = 5,$$

where 10/20 normalize the expected benefit of β_1 to the fact that there are only 10 clients in β_2 (and not 20), and $10 \cdot (90 - 77.5)$ represent the cost of reducing the monthly rate from the average of 90 (in β_2) to the range of 77.5 (in β_1).

Moving from β_2 to β_1 seems attractive. However, we cannot be sure that our estimation for the staying probability (15/22) is accurate. Therefore, we decrease the magnitude of the difference in benefits by a factor of the lower-bound of the confidence interval:

$$w(\beta_2, \beta_1) = \frac{15}{22} - 1.96 \cdot \sqrt{\frac{\frac{15}{22} \cdot \left(1 - \frac{15}{22}\right)}{20}} = 0.478.$$

The effect of this weight leads to the following utility:

$$\text{utility}(\beta_2, \beta_1) = \left(620 \cdot \frac{10}{20} - 180\right) \cdot 0.478 - 10 \cdot (90 - 77.5) = -62.9.$$

Inputs:

- *DT*: a decision tree, which was trained using some decision-tree classification algorithm over a given training set.
- *C*: cost function (see above)
- *B*: benefit function (see above)
- min_value: a threshold for the minimal move utility, in order to be included in the recommendations output.

1. Initialize list_of_recommended_moves to be an empty list of moves
2. Go over all possible pairs of branches, β_1 and β_2, in *DT*:
    2.1. If utility$(\beta_1, \beta_2) \geq$ min_value
        2.1.1. Add O_{12} (the single-branch move from β_1 to β_2), to list_of_recommended_moves
    2.2. If utility$(\beta_2, \beta_1) \geq$ min_value
        2.2.1. Add O_{21} to list_of_recommended_moves
3. Output list_of_recommended_moves

Fig. 3.2 An algorithm for systematically scanning the branches of any given decision tree, and extracting a list of advantageous single-branch moves

That is, with this weighting we will refer to the move as non-advantageous.

In the following section we propose a simple optimization algorithm that receives a decision tree and propose moves based on the utility function.

3.3   An Optimization Algorithm for Proactive Decision Trees

The utility function defined in the previous section provides the hypothetic advantage of acting on observations within a branch in order to change their explaining attributes in a way that will move them to another branch. Based on this function, we suggest a simple algorithm for systematically scanning the branches of any given decision tree and extracting a list of all advantageous single-branch moves. The algorithm is described in Fig. 3.2. The output of the algorithm consists of all the moves with corresponding utility, which is greater than some threshold. This threshold may be set to zero.

Notice that as long as the attribute change cost and benefit functions maintain the triangular equality (that is, the utility of moving from β_1 to β_2 equals the sum of utilities moving from β_1 to β_k and then from β_k to β_2, for all k), the single-branch moves in the output list of recommended moves can be used in any order (and still result in the same total gained utility). From our experience with real life examples,

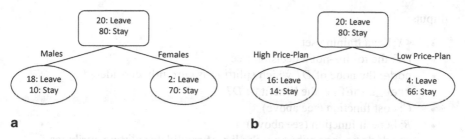

Fig. 3.3 Explaining customers churn. **a** Splitting by customer's gender, and **b** splitting by customer's price plan

(see Chap. 4), it is often the case that the triangular equality is indeed maintained (or almost maintained).

The algorithm in Fig. 3.1 can receive any decision tree that was trained over a classification training set. Existing decision tree algorithms aim for classification accuracy as the optimization criterion. As a result, these algorithms search for splitting rules that contribute to node homogeneity. In the following section we argue that in proactive data mining, the pursuit of accuracy might be misleading. Instead, we propose a novel splitting rule based on the utility function.

3.4   The Maximal-Utility Splitting Criterion

Classification accuracy is the most common criterion for evaluating the quality of classification algorithms. While classification accuracy is important, since we are in fact seeking a classification model that closely simulates reality, excessive emphasis on accuracy might endanger the overall capability of the model. For example, a decision tree that was trained to yield maximal accuracy might use explaining attributes whose values cannot be changed, where there might be surrogate splits, only slightly less accurate, by attributes with values that can be changed easily. This pitfall is demonstrated in Example 3.3.

Example 3.3 Let us consider two possible splits for the problem of churn prediction: (a) splitting according to customers' gender, and (b) splitting according to customers' price plan (which for simplicity can be either high or low). It is possible that the split by customers' gender looks as shown in Fig. 3.3a. This split results in a 88 % accuracy measure. If Fig. 3.3a describes reality, the customers' gender appears to explain the churning well. However, in proactive data-mining we seek for actions. The company, of course is unable to affect the customers' gender. It might be the case that that the split by customers' price-plan looks as shown in Fig. 3.3b. This split results in only 82 % of accuracy (inferior to the split by customers' gender), however, there is a reason to believe that the company can act to reduce the price-plan for the high-paying customers, which in turn may reduce the churning probability.

In order to produce decision trees with a high potential for advantageous moves, we propose a novel splitting criterion, termed the *maximal utility splitting criterion*:

Inputs:

- $<X;Y>$: a training set
- DT: the to-this-point decision tree
- node: the node of DT which splitting is currently considered
- tree_benefit: the benefit of DT
- C: cost function (see above)
- B: benefit function (see above)
- candidate_attrubutes: the list of candidate splitting attributes

1. splitting_attribute = NULL
2. max_utility = total utility that can be achieved from single-branch moves on DT
3. for every attibute in candidate_attrubutes
    3.1. evaluate splitting node according to attribute
    3.2. if the total utility that can be achieved from single-branch moves on the tree after that split exceeds max_utility
        3.2.1. max_utility = the total utility that can be achieved from single-branch moves on the tree after that split
        3.2.2. splitting_attribute = attribute
4. Output splitting_attribute

Fig. 3.4 The maximal utility splitting criterion

splitting according to the values of the explaining attribute in order to maximize the potential total utility that can be gained from the tree. This splitting criterion is described in details, in Fig. 3.4. We use Example 3.4 to illustrate this splitting rule and demonstrate the properties and potential usage of the methods in this chapter.

Example 3.4 This example uses a toy dataset of observations extracted from the activities of a wireless operator's 160 customers—68 left the company; 92 remained. The customers are described by three explaining attributes: A1—describing the customer's package, which can take the values: 'Data', 'Voice' and 'Data&Voice'; A2—describing the customer's sex, which is either 'Female' or 'Male'; and A3—describing the customer's monthly rate in US dollars, with the following possible values: 75, 80, 85, 90 and 95. Table 3.2 describes the empirical joint distribution of the explaining and target attributes.

In our illustration we first generate a decision-tree for predicting the target (Did Churn?) attribute, using the well-known J48 implementation of Weka. The output decision tree is described in Fig. 3.5. J48 builds the tree without considering either the benefit of a customer staying or leaving or the costs of potential changes in the values of the explaining attributes. Notice that the most prominent attribute in this tree is the customer's sex, which clearly the company cannot change.

Table 3.2 The empirical joint distribution of the explaining and the target attributes for example 3.4

Package	Sex	Monthly rate	Did churn?	Number of observations
Data	Female	70	Stay	18
Data	Female	70	Leave	2
Data	Male	70	Leave	20
Data	Female	75	Stay	18
Data	Female	75	Leave	2
Data	Male	75	Stay	8
Data	Male	75	Leave	12
Voice	Female	80	Leave	4
Voice	Male	80	Leave	6
Voice	Male	80	Stay	14
Voice	Female	85	Stay	2
Voice	Female	85	Leave	4
Voice	Male	85	Stay	6
Voice	Male	85	Leave	4
Data&Voice	Female	90	Stay	10
Data&Voice	Female	95	Stay	6
Data&Voice	Female	95	Leave	4
Data&Voice	Male	90	Stay	6
Data&Voice	Male	90	Leave	4
Data&Voice	Male	95	Stay	4
Data&Voice	Male	95	Leave	6

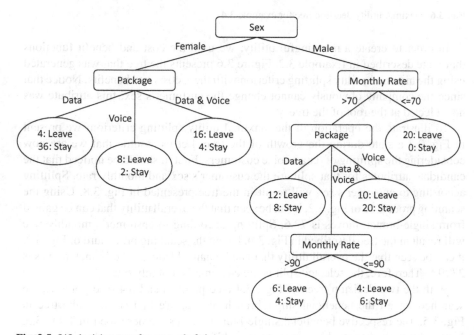

Fig. 3.5 J48 decision tree for example 3.4

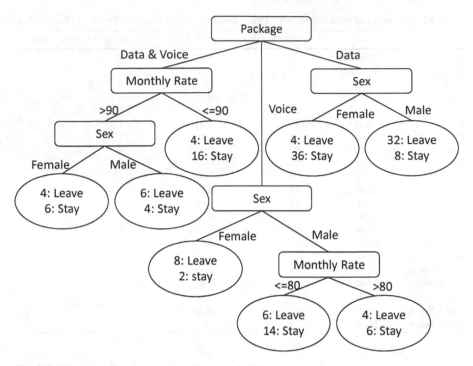

Fig. 3.6 Maximal utility decision tree for example 3.4

In order to create a meaningful utility, we used the cost and benefit functions that were described in Example 3.2. Figure 3.6 presents the tree that was generated using the maximal-utility splitting criterion with these costs and benefits. Notice that since the company obviously cannot change the customer's sex, this attribute was not selected at the root of the tree.

To illustrate the operation of the maximal utility splitting criterion, we present in Fig. 3.7 a state during the growth of the tree. Let us assume that we are now considering the splitting of data&voice customers. First, it should be noticed that the candidate attributes for that split are the customer's sex and monthly rate. Splitting according to customer's sex will result in the tree presented in Fig. 3.8. Using the scanning procedure of Fig. 3.2, it can be seen that the overall utility that can be gained from single-branch moves is 72.6. Splitting according to customer's monthly rate will result in the tree described in Fig. 3.9. Using the scanning procedure of Fig. 3.2, it can be seen that the overall utility that can be gained from single-branch moves is 2799.5. Therefore, the selected split is the customer's monthly rate.

With the two decision trees of Figs. 3.4 (see point 3) and 3.4 (see point 4), we searched for all the beneficial single-branch moves. We first scanned the tree in Fig. 3.5. The respective beneficial single-branch moves are described in Table 3.3. Notice that the most valuable moves strive to shift customers from a node with a churn rate of 80 % to a node with churn rate of merely 10 %. The overall benefit of the tree in Fig. 3.5 is: 2020. After implementing all the valuable moves of Table 3.3, we end up with a tree with an overall benefit of: 3000. We then scanned the tree in Fig. 3.6

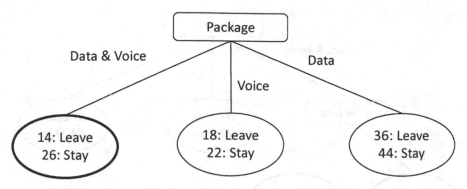

Fig. 3.7 The maximal utility splitting criterion

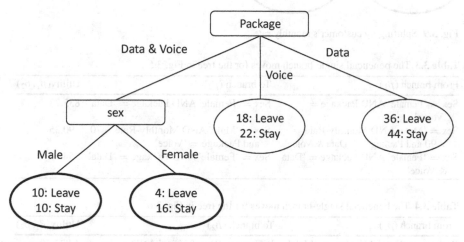

Fig. 3.8 Splitting by customer's sex

(which was constructed using the maximal utility splitting criterion, along with the meaningful cost and benefit functions described above). The respective beneficial single-branch moves are described in Table 3.4. Although the tree in Fig. 3.6 has a benefit of 2020 as in Figure 3.4.3, after implementing all the beneficial single-branch moves, the overall pessimistic benefit raises to 5435.11 (which is significantly higher than 3000). Moreover, it can be seen that while the beneficial single-branch move in Table 3.3 are mainly refinements of the tree (they are moves across relatively low splits of the tree) the moves in Table 3.4 change the basic tree significantly.

3.5   Chapter Summary

In this chapter, two decision tree based implementations for proactive data mining were proposed. We began the chapter by discussing the advantages of decision tree algorithms, such as their simplicity and the explicit description of the dependencies of

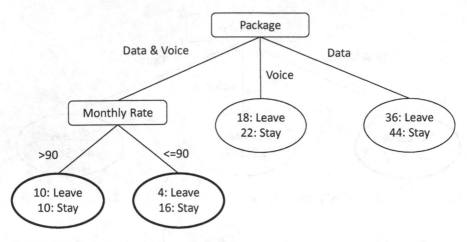

Fig. 3.9 Splitting by customer's monthly rate

Table 3.3 The beneficial single-branch moves for the tree in Fig. 3.5

From branch (β_1)	To branch (β_2)	Utility(β_1, β_2)
Sex = 'Female' AND Package = 'Voice'	Sex = 'Female' AND Package = 'Data'	835.23
Sex = 'Male' AND Monthly-Rate > 90 and Package = 'Data & Voice'	Sex = 'Male' AND Monthly-Rate > 70 and Package = 'Voice'	90.35
Sex = 'Female' AND Package = 'Data & Voice'	Sex = 'Female' AND Package = 'Data'	54.64

Table 3.4 The beneficial single-branch moves for the tree in Fig. 3.6

From branch (β_1)	To branch (β_2)	Utility(β_1, β_2)
Package = 'Data' AND Sex = 'Male'	Package = 'Data&Voice' AND 'Monthly-Rate' $\leq$ 90	1,873.70
Package = 'Voice' AND Sex = 'Female' AND Monthly-Rate > 70	Package = 'Data' AND Sex = 'Female'	835.23
Package = 'Data&Voice' AND 'Monthly-Rate > 90 AND Sex = 'Male'	Package = 'Data&Voice' AND Monthly-Rate $\leq$ 90	380.01
Package = 'Data&Voice' AND 'Monthly-Rate > 90 AND Sex = 'Female'	Package = 'Data' AND Sex = 'Female'	304.43
Package = 'Voice' AND Sex = 'Male' and Monthly-Rate > 80	Package = 'Voice' AND Sex = 'Male' and Monthly-Rate $\leq$ 80	21.65

the target attribute, in the explaining attributes. We then described our proposed utility measure in detail. We proposed a simple algorithm for scanning a given decision tree for pairs of branches while looking for moves with a sufficiently high corresponding utility. Finally, we proposed a novel splitting rule for decision tree algorithms in which the split occurs according to the values of the explaining attribute that maximizes the utility of the resulting tree.

References

Ben-Shimon D, Tsikinovsky A, Rokach L, Meisles A, Shani G, Naamani L (2007) Recommender system from personal social networks. In: Wegrzyn-Wolska K , Szczepaniak P. (eds), Advances in Intelligent Web Mastering. Advances in Soft Computing, vol 43, Springer Berlin Heidelberg, pp 47–55

Kisilevich S, Rokach L, Elovici Y, Shapira B (2010) Efficient multidimensional suppression for k-anonymity. IEEE Trans Knowl Data Eng 22(3):334–347

Matatov N, Rokach L, Maimon O (2010) Privacy-preserving data mining: a feature set partitioning approach. Inf Sci 180(14):2696–2720

Menahem E, Rokach L, Elovici Y (2009) Troika–An improved stacking schema for classification tasks. Inf Sci 179(24):4097–4122

Rokach L (2009) Collective-agreement-based pruning of ensembles. Comput Stat Data Anal 53(4):1015–1026

Chapter 4
Proactive Data Mining in the Real World: Case Studies

This chapter presents two real world implementations of the proactive data mining, using decision trees from two different sectors: cellular services (Sect. 4.1) and security (Sect. 4.2). Using actual datasets, we address real problems that two companies in these business areas face.

4.1 Proactive Data Mining in a Cellular Service Provider

In an extremely competitive and saturated environment, managing customer churning is extreme important to the survival of any wireless operator (Hung et al. 2006; Arbel and Rokach 2006). In Israel, the telecom market has recently gone through a significant reform that has enabled new competitors to enter the market. At the same time clients gained the opportunity to change their wireless operator almost cost-free. Consequently, the reform also led to a drastic drop in revenues and profits among virtually all Israeli wireless operators. In the last year alone, five new wireless operators joined the fray, each offering extremely low rates. The pre-reform market leaders who were now losing customers to the new competitors, began to downsize their businesses and look for ways to maintain profitability. Data mining became one of the primary technologies that the operators turned in hopes of salvaging the situation.

4.1.1 The Data Mining Problem for the Wireless Company

As part of the reform, one of the leading wireless operators in Israel was granted an additional operator's license. To counter its new, low-price competitors, while maintaining its existing business and positioning itself as a prestigious and high-end service for premium paying customers, the operator formulated the idea of creating a totally new cellular company that would offer discount rates (a flat fee of 90NIS/month). However, the new license that the cellular company was able to obtain did not come free; the government attached several conditions. One condition

H. Dahan et al., *Proactive Data Mining with Decision Trees*,
SpringerBriefs in Electrical and Computer Engineering,
DOI 10.1007/978-1-4939-0539-3_4, © The Author(s) 2014

required that during the first 3 years of operation, the new company must see to it that at least 25 % of its customers reside on the country's periphery. That is, for every three new customers it recruits (or mobilizes from the existing company) from the center of the country, at least one customer must live on the periphery. From the government's perspective, this condition would force the new wireless operators to invest and develop a wireless infrastructure not just in the heavily developed and populated country center, but also in the more rural, less densely populated regions. For the wireless operators, this requirement, posed a serious obstacle to its rate of growth.

To meet the government conditions, while maintaining a rapid growth rate, the wireless operator decided to migrate some of its periphery customers (from the high service, premium company) to its newly created discount company. Obviously, the objective was to migrate the periphery customers in such a way that the benefit generated from the migration to the new discount operation would outweigh the cost of migration and the loss of income from these premium paying customers. By moving these customers from the existing wireless operation to the new discount wireless operation, the company would eliminate the limitation on its growth rate by meeting the government's condition. At the same time it would be minimizing income loss to its existing wireless operation.

The problem at hand is not a traditional churn dilemma: "Who are the most probable customers that will leave the company?" The problem is an optimization task: "Which customers (among those residing on the periphery) would it be best to migrate to the new discount business?" In seeking a solution to this task, various parameters had to be considered, specifically those that indicated which group of customers would generate the best potential utility for the organization. That is to say, the operator had to evaluate the importance of some features to the customer, potential loss of income versus the potential utility it generates in its new discount wireless operation

4.1.2   The Wireless Dataset

Of the 240,000 customer records in the dataset we received from the operator, 42,750 lived on the periphery. Since our objective was to migrate customers to the discount operation from the existing company, we required samples from the new discount customers. Therefore, to the 42,750 records, we added 1,250 records from the new discount operation's customer dataset. Consequently, the combined dataset contained 45,000 records of two types of customers: existing premium paying customers and new discount customers. Of the customers in the combined dataset 95.7 % were labeled as "stay" and 4.3 % as "leave". Although the operator's real dataset consisted of many attributes, for the sake of simplicity we used only the following eight most relevant attributes:

(i) Customer-Type—A new attribute we added to the combined dataset to differentiate between existing and new discount customers.
(ii) Customer-Seniority—The time elapsed since the customer joined the operator (in months). This attribute is an indicator of customer loyalty to the operator.
(iii) Monthly-Rate—The average monthly payment (over the last 3 months) the customer paid the operator. This is an important parameter since migrating a customer with a high monthly payment will lead to loss in income for the operator.
(iv) Usage-Within-Network—The average amount of minutes (for the last 3 months) the customer talks with other customers of the operator (within the operator's network). The use of this feature is free of charge.
(v) Usage-Outside-Network—The amount of minutes (per month) the customer talks with people outside the operator's network.
(vi) Minutes-of-Use—The total amount of air time (number of minutes) the user talked (per month). This attribute includes minutes within the operator's network, outside the operator's network, incoming calls, etc.
(vii) Total-KB-Data-Usage—The amount of kilobytes the customer used.

In addition to the above explaining attributes, the combined dataset contained a target attribute termed "Churn" which had two possible values "Leave" or "Stay".

4.1.3 Attribute Discretization

Except for the new "Customer-Type" attribute which is of a nominal type; all other explaining attributes in the combined dataset are numeric. To reduce the magnitude of the data mining model and generate readable decision trees, we preprocessed the combined dataset and discretized its attributes. For example, the value of the "Usage-Within-Network" attribute varied from 0 to 7000 [minutes]. From the operator's point of view, there is no major difference between a customer using the network for 5 min or 15 min or between those using it for 30 or 35 min. Therefore, we re-labeled all attribute values into fewer preset groups. For instance, all values between 0 and 25 were set as 25, those between 26 and 50 were revalued as 50, etc. Similarly, the attribute "Customer-Seniority" values varied from 0 to 120 [months]. Therefore, we divided the values into groups based on the number of customers in each group.

An important feature that our algorithm offers is that it allows for usage of what we call "silent attributes". These are attributes that are part of the dataset and are used in the calculations (i.e., calculating attribute value changing cost and/or benefit function) during the construction of the data mining model but are not part of the decision tree structure. This feature allow us to use normalized attributes for constructing the decision tree while maintaining the original attribute values for utility calculations such as "silent attributes". In this way, we get a very readable decision tree without losing the importance of the varied attribute values. For example, the values of the customer's monthly payment vary widely. We turned the original "Monthly-Rate" attribute into a "silent attribute" and added a new normalized version of the "Monthly-Rate" attribute. We grouped the values of the new attribute into batches. For instance, for all customers with the original version of the "Monthly-Rate" (i.e., the "silent attribute") with values between 0 and 50, we set the value of the new version of the "Monthly-Rate" attribute to 25.Those between 51 and 100 were set as 75 and so on. When we calculated the potential utility of the different moves, we used the original value from the "silent attribute" version of the "Monthly-Rate", but for construction of the decision tree we used the new formalized version. This "silent attribute" mechanism allowed us to generate simpler decision tree models while providing accurate values during calculations of the utility or cost functions.

4.1.4 Additional Environment and Problem Knowledge for the Wireless Company

As indicated above, one of the major differences between proactive and passive data mining methods is the capability of the proactive method to consider additional knowledge during decision model construction and the optimization phase. As a result, the constructed models and actions that the Maximal Utility algorithm suggests are tailored and more relevant to the problem at hand.

In our case, in addition to the dataset indicated above, the wireless operator has other knowledge that affects the model construction and optimization processes:

(i) The existing wireless operator has a unique feature that allows customers to talk almost freely within the operator's network. This feature is best for customers who have more than one account with the operator as well as for companies that provide their employees with cell phones. This feature enables operator to track the time that customers within this network use. Companies that extensively use this feature may not be good candidates to migrate to the new discount operator (which does not offer this feature).

(ii) The monthly income it receives from its existing customers is of great value. There is no business sense in migrating to a discount operator a customer who generates substantial monthly income and profit. Nor does it make sense to lose income by offering a lower monthly payment, unless, of course, the chances that this customer might leave are high due to the new market reform.

(iii) The wireless operator knows that it generates monthly profit of at least 10NIS for each new non-periphery customer it adds to its new discount operator business.

In addition to the above knowledge, to construct a model that maximizes the utility function set by the organization, the proactive model must also encompass all the costs and benefits associated with possible value changes in any of the attributes. The following section describes these costs and benefits for the wireless operator.

Table 4.1 Cost matrix for Customer-Type

	Discount	Periphery
Discount	0	∞
Periphery	− 360	0

Table 4.2 (High) Cost matrix for Customer-Seniority

Changing to	Cost function
< Current "Customer-Seniority"	(5 * (Current Customer-Seniority—Target Customer-Seniority))
> Current "Customer-Seniority"	∞

4.1.4.1 Cost Matrices

It is important to carefully consider how to set up the different cost and benefit matrices for all possible attribute values since the way we set these matrices can cause the algorithm to search for potential migration actions in one direction rather than another. Since the operator's objective is to find the potentially most beneficial actions that will lead periphery customers to migrate to the new discount operation, we have to set the matrices in such a way that migration to the discount is preferred and not discouraged.

Given the operator's stated objectives, in order to encourage migration from periphery customers into the discount operation, we set the cost matrix of the "Customer-Type" as shown in Table 4.1.

We set an infinite cost for changing a customer type from a discount into a periphery customer to discourage such moves. On the other hand, the cost of changing the customer type from periphery into discount generates income of 360NIS to the operator. That is, 90NIS for the migrated periphery customer flat fee payment as new discount customer, plus 3 × 90NIS for the other three discount customers the operator is allowed to have according to the government condition (25 % periphery customers' vs. 75 % non-periphery customers).

We argue that the exact values in the matrices are insignificant. The important thing is to set values that encourage the migration of the periphery customers into discount. This is done by setting a smaller value in the cost of changing the value of the "Customer-Type" attribute from "Periphery" into "Discount". In this way, the algorithm will give preference to actions that suggest moving a customer from periphery into discount, which is exactly what we are looking for. From these potential actions, we will choose those with the best utility for the operator.

The "Customer-Seniority" attribute indicates the number of months this customer has been with the operator. The operator's marketing experts are not clear whether this factor is stronger than the "Monthly-Rate" in the customer's decision to migrate. Therefore we provided two scenarios: one with high cost matrices change values for "Customer-Seniority" and "Usage-Within-Network" attributes and one will lower values. For the first scenario, we set the cost matrix for "Customer-Seniority" as shown in Table 4.2.

Table 4.3 Cost matrix for Monthly-Rate

Changing to	Cost function
< Current "Monthly-Rate"	(Current "Monthly-Rate"—Target "Monthly-Rate")
> Current "Monthly-Rate"	∞

Table 4.4 (High) Cost matrix for Usage-Within-Network

Changing to	Cost function
< Current "Usage-Within-Network"	(current "Usage-Within-Network"—target "Usage-Within-Network")
> Current "Usage-Within-Network"	0

The matrix above indicates that the cost of changing a customer is also dependent on seniority. That is, "moving" a customer with higher seniority into the discount operation is more "expensive" than moving a customer with lower seniority. We set an infinite value as the cost of changing the value of the "Customer-Seniority" attribute of customers with lower seniority to customers with higher seniority. This is done in order to discourage discount customers with lower seniority becoming periphery customers with higher seniority. On the other hand, when the system checks for potential migrations from periphery customers into discount customers (i.e., from customers with higher seniority into customers with lower seniority), the cost depends on the seniority of the customer. The cost function is an expression that multiplies by five the difference between the current customer's seniority (the value of "Customer-Seniority" attribute of the customer(s) on the source path) and the target customer's seniority (the value of the "Customer-Seniority" attribute of the customer(s) on the target path). Note, that when there is more than one customer instance on the source or the target paths, the system calculates the average "Customer-Seniority" of all instances.

The "Monthly-Rate" attribute indicates the monthly income the operator receives from each customer. The cost matrix for this attribute is described in Table 4.3:

The matrix above indicates that the cost of changing a customer with a higher "Monthly-Rate" value into a lower "Monthly-Rate" value is the difference between values of the source and target attributes. For example, leading a periphery customer who pays 200NIS to pay only 90NIS is: $200 - 90 = 110$. On the other hand, the matrix discourages any increase in the monthly payment. That is, we assume that customers will not object to their monthly rate reduction but will resist any increase in monthly payments (which also may be forbidden by the regulations).

The "Usage-Within-Network" attribute indicates the customer's monthly usage (in minutes) of the operator's free network. Some of the operator's marketing experts believe that this attribute provides an important indicator of the customer's tendency to "move" to the discount operator or any other operator since this feature is unique to this operator. A customer that has a high "Usage-Within-Network" value is less likely to move than a customer with a lower value. We set the cost matrix of this attribute is described in Table 4.4.

Table 4.5 Cost matrix for Usage-Outside-Network

Changing to	Cost function
< Current "Usage-Outside-Network"	(current "Usage-Outside-Network"—target "Usage-Outside-Network")
> Current "Usage-Outside-Network"	∞

Table 4.6 Cost matrix for Minutes-of-Use

Changing to	Cost function
< Current "Minutes-of-Use"	(current "Minutes-of-Use"—target "Minutes-of-Use")
> Current "Minutes-of-Use"	∞

Table 4.7 Cost matrix for Total-KB-Data-Usage

Changing to	Cost function
< Current "Total-KB-Data-Usage"	(current "Total-KB-Data-Usage"—target "Total-KB-Data-Usage")
> Current "Total-KB-Data-Usage"	∞

The matrix above indicates that the cost of changing a customer with a higher "Usage-Within-Network" value into a lower "Usage-Within-Network" value is the difference between the values of the source and target attribute. For example, changing a periphery customer who uses 500 min of the operator's free network into a discount customer who talks freely for 300 min (i.e., the flat fee for the new discount customers includes a basic package of minutes) is: $500 - 300 = 200$. On the other hand, the cost of not using the free operator network is 0. Similarly, the cost matrices for the attributes "Usage-Outside-Network", "Minutes-of-Use" and "Total-KB-Data-Usage" are described in Tables 4.5, 4.6 and 4.7, respectively.

It is important to note that the cost and the benefit functions are calculated for all instances according to the source and target paths under consideration in the decision tree (see definitions in Chap. 2 and 3). If there is more than single instance on a path, the average of all said instances is calculated and used in these functions.

4.1.4.2 Benefit Matrix

The benefit matrix represents the positive or negative gain associated with each target attribute value. The dataset includes a churning indication. Although the problem facing the wireless operator is not churn prediction, we use the churning indication to assist in determining the benefits that the firm gains from its clients. When we consider moving periphery residing customers to the discount operation, we take into consideration the benefit loss (or gain) by reducing the new income (as a discount customer) from the old income (as an existing, periphery-residing customer). Our objective is to find the actions that are still beneficial to the operator despite the

Table 4.8 Benefit matrix for the wireless company	Status	Benefit
	Stay	(Monthly-Rate)
	Leave	(− Monthly-Rate)

change in income. The benefit matrix is set to represent the outcome from the client. The benefit the operator gains from a staying (non-churning) client is set at his or her monthly rate. The benefit from a leaving (churning) client is set at minus that monthly rate, as described in Table 4.8.

Note that even with no other environmental or problem knowledge, with only the benefit matrix, the proactive data mining method constructs models that provide more relevant and beneficial information than the passive zero-one loss data mining methods. For example, for the traditional churn problem, a zero-one loss classification model will indicate that customers that are more likely to leave or stay regardless of their contribution to the organization income. That is, for the zero-one loss passive method, a customer who pays $ 1,000/month is equivalent to one paying $ 10/month. This could dilute the focus of the organization and lead it to chase after low paying customers. On the other hand, the proactive model with a benefit matrix will place more weight on high paying customers, leading the organization to focus mainly on those customers who contribute most to its income.

4.1.5   Passive Classification Model for the Wireless Company

Given the above dataset, the traditional, passive classification methods might help the wireless company to find the characteristics of the subgroups of potential customers most likely to stay or leave. This information can help the company determine which of its existing customers are the best candidates to "move" to its new discount business and which should remain with the existing premium business. As indicated above, in expanding its discount business, the company must meet the government requirement that 25 % of its discount customers must reside on the periphery. To meet this requirement, the company benefits by offering to move to its new discount business those of its customers who are low paying, reside on the periphery and most likely to leave. In this way, the operator keeps those most likely to stay as premium paying customers.

The passive data mining models use only the dataset as input. We generated a passive model using a J48 implementation of Weka (Waikato Environment for Knowledge Analysis 1999–2011) over the wireless dataset described above. We later compared it with our proactive generated model. Figure 4.1 describes the J48 generated decision model for the wireless company.

- Monthly-Payment <= 50
  - Usage-Outside-Network <= 0: Stay (2169/298)
    - Usage-Outside-Network > 0
      - Monthly-Payment <= 25
        - Customer-Seniority <= 6
          - Usage-Outside-Network <= 10
            - Total-KB-Data-Usage <= 20: Stay (9/3)
            - Total-KB-Data-Usage > 20: Leave (2/0)
          - Usage-Outside-Network > 10: Stay (3/0)
        - Customer-Seniority > 6
          - Usage-Outside-Network <= 30: Leave (24/0)
          - Usage-Outside-Network > 30: Stay (5/1)
      - Monthly-Payment > 25
        - Customer-Seniority <= 9: Stay (85/10)
        - Customer-Seniority > 9
          - Customer-Seniority <= 12: Leave (16/5)
          - Customer-Seniority > 12: Stay (25/10)
- Monthly-Payment > 50: Stay (42662/1569)

Number of leaves: 10
Number of attributes: 18
Total Benefit: 9313363
Tree Accuracy: 95.72%
Correctly Classified Rows: 43104
Incorrectly Classified Rows: 1896

Fig. 4.1 J48 generated model for the wireless company

4.1.6   *Maximal Utility Generated Model for the Wireless Company*

The cost and benefit matrices represent the problem specific and environmental knowledge that must be incorporated into a model for it to be beneficial and applicable. Using our Maximal Utility splitting criterion along with the cost and benefit functions over a combined dataset of 45,000 customer records described above; we produced a decision tree whose output is described in Fig. 4.2. It should be noted that by setting special "incentives" in the cost functions for the "Customer-Type" attribute, this attribute was chosen as the root node of the decision tree. In addition, since all "discount" customers are new to the company, they were set with similar characteristics (i.e., the main incentive is the flat, low, monthly fee). Accordingly, these discount customers are represented in the resulting tree by a single path with a single leaf. The initial, overall, pessimistic benefit of the tree is: $TB(DT) = 9,313,363$. The accuracy of the tree as measured by a 10-fold cross validation procedure is 95.72 % (which is identical to the accuracy of the passive J48 generated decision tree).

- Customer-Type = "Discount": Stay (1250/0)
- Customer-Type = "Periphery"
  - Usage-Within-Network > 500: Stay (239/11)
  - Usage-Within-Network <= 500
    - Usage-Within-Network > 150
      - Usage-Within-Network > 400
        - Total-KB-Data-Usage > 250: Leave (1/0)
        - Total-KB-Data-Usage <= 250: Stay (128/7)
      - Usage-Within-Network <= 400
        - Monthly-Payment > 325
          - Total-Air-Time > 350: Stay (570/18)
          - Total-Air-Time <= 350: Stay (9/1)
        - Monthly-Payment <= 325
          - Total-KB-Data-Usage > 250: Stay (17/1)
          - Total-KB-Data-Usage <= 250: Stay
    - Usage-Within-Network <= 150
      - Usage-Within-Network > 100
        - Monthly-Payment > 300
          - Monthly-Payment > 325: Stay (392/10)
          - Monthly-Payment <= 325: Stay (94/1)
        - Monthly-Payment <= 300
          - Total-KB-Data-Usage > 300: Stay (3/0)
          - Total-KB-Data-Usage <= 300
            - Monthly-Payment > 100: Stay (779/38)
            - Monthly-Payment <= 100: Stay (13/1)
      - Usage-Within-Network <= 100
        - Usage-Within-Network > 50
          - Monthly-Payment > 325
            - Total-Air-Time > 450: Stay (631/18)
            - Total-Air-Time <= 450: Stay (133/4)
          - Monthly-Payment <= 325
            - Monthly-Payment > 175: Stay (1646/60)
            - Monthly-Payment <= 175: Stay (384/14)
        - Usage-Within-Network <= 50
          - Usage-Within-Network > 25
            - Total-Air-Time > 50
              - Monthly-Payment > 175: Stay (2384/81)
              - Monthly-Payment <= 175: Stay (482/35)
            - Total-Air-Time <= 50: Stay (96/4)
          - Usage-Within-Network <= 25
            - Total-Air-Time > 150: Stay (8747/326)
            - Total-Air-Time <= 150
              - Monthly-Payment > 125: Stay (17739/610)
              - Monthly-Payment <= 125: Stay (8157/638)

Number of leaves: 23
Number of attributes: 44
Total Benefit: 9313363
Tree Accuracy: 95.72%
Correctly Classified Rows: 43073
Incorrectly Classified Rows: 1927

Fig. 4.2 Maximal Utility generated model for the wireless company (with strong emphasize on Customer-Seniority and Usage-Within-Network attributes)

- Customer-Type = "Discount": Stay (1250/0)
- Customer-Type = "Periphery"
  - Total-Air-Time > 2000
    - Customer-Seniority > 6
      - Customer-Seniority > 12: Stay (58/2)
      - Customer-Seniority <= 12: Stay (123/6)
    - Customer-Seniority <= 6: Stay (89/2)
  - Total-Air-Time <= 2000
    - Total-Air-Time > 1950: Stay (33/1)
    - Total-Air-Time <= 1950
      - Customer-Seniority > 9
        - Total-Air-Time > 1750: Stay (43/0)
        - Total-Air-Time <= 1750
          - Monthly-Payment > 250: Stay (3720/123)
          - Monthly-Payment <= 250: Stay (12609/660)
      - Customer-Seniority <= 9
        - Total-Air-Time > 1450: Stay (342/10)
        - Total-Air-Time <= 1450
          - Monthly-Payment > 225: Stay (10474/293)
          - Monthly-Payment <= 225: Stay (16259/831)

Number of leaves: 11
Number of attributes: 20
Total Benefit: 9313363
Tree Accuracy: 95.72%
Correctly Classified Rows: 43072
Incorrectly Classified Rows: 1928

Fig. 4.3 Maximal Utility generated model for the wireless company (with moderate emphasize on Customer-Seniority and Usage-Within-Network attributes)

So far we have generated two decision trees. One is based on the passive data mining method J48 (Fig. 4.1) while the second uses our proactive Maximal Utility (Fig. 4.2). Note that T both decision models have the same initial total benefit of $TB(DT) = 9,313,363$. This reflects the current benefit of all 45,000 customers in the dataset. The company's objective is to find the potential beneficial actions or moves that maximize the utility function set by the company (i.e., income). To realize this objective, we need to perform the second phase of our proactive method—run the decision tree optimization algorithm over both decision models (Figs. 4.2 and 4.3) and compare the potential utility gains that are generated.

Table 4.9 Optimized algorithm generated action list over the J48 model for the wireless company (with high cost values for Customer-Seniority and Usage-Within-Network attributes)

From branch (β_i)	To branch (β_j)	Utility (β_i, β_j)
Monthly-Payment ≤ 50 AND Usage-Outside-Network > 0 AND Monthly-Payment > 25 AND Customer-Seniority > 12	Monthly-Payment ≤ 50 AND Usage-Outside-Network ≤ 0	333.73

4.1.7 Optimization Algorithm over the J48 Generated Model for the Wireless Company

To generate from the J48 model the potentially beneficial moves or actions that will presumably benefit the operator, we perform the second phase of the utility maximization method (i.e., running the optimization algorithm—see Chap. 2) on the J48 decision tree (Fig. 4.1). The optimization algorithm systematically scans the branches of the decision tree that was produced in the first phase and extracts a list of all the advantageous single-branch moves. Running the optimization algorithm over the J48 decision tree results in only one action that displays small utility gain, as shown in Table 4.9.

Implementing the suggested action over the J48 generated decision tree results in an overall additional utility gain of Utility (β_i, β_j) = 333.73. This represents a negligible total utility gain (i.e., 0.00035 %) over the initial total benefit of the J48 decision tree (Fig. 4.1).

4.1.8 Optimization Algorithm over the Maximal Utility Generated Model for the Wireless Company

Implementing the optimization algorithm on the Maximal Utility decision tree (Fig. 4.2)we find that the list of suggested potential actions (Table 4.10) is much longer than with the passive data mining model. However, for the same of simplicity, in Table 4.10 we listed only those actions that generate a minimum utility gain of 350,000. The Maximal Utility method considers the knowledge that is provided during the construction of the decision tree with the objective of allowing future interventions (i.e., by the optimization algorithm) that maximize the utility function set by the organization. Comparing both suggested action lists, it is clear that given the same dataset and the same additional knowledge (i.e., cost and benefit matrices) the proactive Maximal Utility generated decision tree provides far more options for intervention and a much higher potential utility gain than the passive J48 generated decision tree.

Notice that the first two suggested actions by the system have almost identical characteristics except in regard to the "Monthly-Rate" attribute value. Surprisingly enough, the first suggested action by the system is to move those clients with a "Monthly-Rate" that is greater than 125. Intuitively we would expect that the lower

Table 4.10 Optimized algorithm generated action list over the Maximal Utility model for the wireless company (with high cost values for Customer-Seniority and Usage-Within-Network attributes)

From branch (β_i)	To branch (β_j)	Utility($\beta_i\beta_j$)
Customer-Type = "Periphery" AND Usage-Within-Network $\leq$ 25 AND Minutes-of-Use $\leq$ 150 AND Monthly-Rate > 125	Customer-Type = "Discount"	4,467,977.67
Customer-Type = "Periphery" AND Usage-Within-Network $\leq$ 25 AND Minutes-of-Use $\leq$ 150 AND Monthly-Rate $\leq$ 125	Customer-Type = "Discount"	3,048,879.66
Customer-Type = "Periphery" AND Usage-Within-Network $\leq$ 25 AND Minutes-of-Use > 150	Customer-Type = "Discount"	1,423,800.31
Customer-Type = "Periphery" AND Usage-Within-Network $\leq$ 50 AND Usage-Within-Network > 25 AND Minutes-of-Use > 50 AND Monthly-Rate > 175	Customer-Type = "Discount"	366,571.47

Table 4.11 (Moderate) Cost matrix for Customer-Seniority	Changing to	Cost function
	< Current "Customer-Seniority"	(2 * (Current Customer-Seniority—Target Customer-Seniority))
	> Current "Customer-Seniority"	∞

monthly rate would be more beneficial. However, the system decided that the path with the higher "Monthly-Rate" is the one that will offer the highest utility gain. The reason for this selection is that this path reduces the churning potential (i.e., lower rate of churning) offsetting the additional income loss from the higher "Month-Rate".

Implementing all of the four suggested actions listed in Table 4.10 results in an additional overall utility gain of: Utility $(\beta_i\beta_j) = 9,307,229.11$. This represents a total utility gain of about 100 % over the initial total benefit of the Maximal Utility decision tree (Fig. 4.2).

As indicated, the model generated above was based on cost matrices that place strong value on the role of the "Customer-Seniority" and "Usage-Within-Network" attributes in the user's decision to migrate. These values significantly enhanced the role of "Usage-Within-Network" attribute as reflected in the model generated above (Fig. 4.2).

Another scenario that we checked was the cost matrices for those attributes that place a reduced value in the user's decision to migrate. Using the cost matrices for "Customer-Seniority" and "Usage-Within-Network", we generated another model, as described in Tables 4.11 and 4.12.

These matrices assume that the cost of changing the values of these attributes is half that of the cost in the Maximal Utility model described above (Fig. 4.2). The cost and benefit matrices of the other attributes remained unchanged. In the second decision tree that we produced, the Maximal Utility method is run on the combined dataset of 45,000 customer instances described above. The output is the Maximized

Table 4.12 (Moderate) Cost matrix for Usage-Within-Network	Changing to	Cost function
	< Current "Usage-Within-Network"	(0.5 *(current "Usage-Within-Network"—Target "Usage-Within-Network"))
	> Current "Usage-Within-Network"	0

Utility decision tree described in Fig. 4.3. Notice that the "Customer-Type" attribute is still chosen as the root node of the decision tree. However, the reduced changing cost of the "Customer-Seniority" attribute apparently had more impact on the utility gain as reflected in the generated new Maximal Utility model described in Fig. 4.3. As in the previous model, the initial overall pessimistic benefit of the decision tree is: $TB(DT) = 9,313,363$.

In Fig. 4.3 we see that the second proactive model (with moderate cost values for Customer-Seniority and Usage-Within-Network attributes) also generated four actions with a derived minimum utility gain of at least 350,000. Implementing all four of the actions suggested in Table 4.13 results in an additional overall utility gain of Utility $(\beta_i, \beta_j) = 19,832,235$. This represents a total utility gain of over 200 % for the overall utility gain of the proactive model with the high cost values for Customer-Seniority and Usage-Within-Network attributes (Fig. 4.2). The two generated proactive classification trees demonstrate the importance of environment/problem knowledge to the model and the way it can help in the decision-making process of the organization. The fact that these two models are drastically different (based on the information provided), demonstrates the importance and relevance of domain knowledge to the data mining process. As demonstrated above, the proactive data mining approach can address business problems that cannot be effectively handled or acted upon by traditional, passive data mining methods. More specifically, when considering the distribution of the input observations as changeable, business users can evaluate the outcome of actions that may impact future distribution. The domain knowledge provides the basis for measuring these outcomes. This case study demonstrates how the proactive data mining approach can provide a unique competitive value for a wireless operator in a stormy business environment.

Executing the optimization algorithm over the Maximal Utility decision tree in Fig. 4.3 (with a minimum utility gain of 350,000), results in the following action list:

4.2   The Security Company Case

Crime in recent years has risen considerably and nervous citizens are racing to acquire anti-burglary electronic systems to protect their property. In meeting the demand, security companies are fiercely competing to acquire customers by offering a variety of security systems to protect private homes as well as businesses.

These companies have invested heavily in establishing security centers that provide state-of-the-art surveillance services for their customers. To recover their huge

Table 4.13 Optimized algorithm generated action list over the Maximal Utility decision tree for the wireless company (with moderate cost values for Customer-Seniority and Usage-Within-Network attributes)

From branch (β_i)	To branch (β_j)	Utility (β_i, β_j)
Customer-Type = "Periphery" AND Minutes-of-Use $\leq$ 1450 AND Customer-Seniority $\leq$ 9 AND Monthly-Payment $\leq$ 225	Customer-Type = "Discount"	4,973,133.25
Customer-Type = "Periphery" AND Minutes-of-Use $\leq$ 1750 AND Customer-Seniority > 9 AND Monthly-Payment $\leq$ 250	Customer-Type = "Discount"	3,769,588.42
Customer-Type = "Periphery" AND Minutes-of-Use $\leq$ 1450 AND Customer-Seniority $\leq$ 9 AND Monthly-Payment > 225	Customer-Type = "Discount"	1,373,489.32
Customer-Type = "Periphery" AND Minutes-of-Use $\leq$ 1750 AND Customer-Seniority > 9 AND Monthly-Payment > 250	Customer-Type = "Discount"	402,661.02

investments and achieve a reasonable level of profitability, the security companies are trying to sign up as many customers as possible. Given the high initial investment by the customers themselves in the surveillance equipment that will be placed in their homes, the security companies believe that once a customer has been signed up, it is very unlikely that she or he will leave anytime soon. Thus competition among the security companies for each customer is very strong.

The security companies have been collecting information about potential customers for years. Every sales call and every sale visit to potential customers are documented. The challenge for the companies is to utilize this data along with their vast experience to sign up more customers and to retain existing customers from leaving and moving to their competitors. In this highly competitive market it is not surprise that some security companies are utilizing data mining techniques to cope with the competition.

4.2.1   The Data Mining Problem for the Security Company

The security company we are considering reaches potential customers either by cold calling or through references from existing customers. Due to the strong competition among the security companies, it is important to maximize the number of signed-up customers in relation to the number of potential customers. Given its limited resources and business constraints (Arbel and Rokach 2006), the security company is looking for ways (actions) to improve its chances to sign up as many customers as possible from among potential customers. For example, from past experience the security company learned that the chances of signing a potential customer are much higher when current customers refer the potential customer to the company. Therefore, the company looks for ways to encourage its current customers to refer new potential customers. However, in order to ensure profitability, the company wants to know what limits (in terms of cost) should be placed on the incentives it offers current customers for their referrals.

Traditional classification methods predict the chances of each customer or group of customers subscribing to the company's services. However, these methods do not consider the associated costs or benefit of each potential customer nor do they direct the company towards actions that might improve the future customer distribution (i.e., impact the behavior) of the potential customers to its benefit. The goal of the company is find the information about the course of action it should follow in order to improve its profits, given the constraints under which it operates. The problem at hand is not a traditional prediction problem of "Who are the customers most likely to sign?" The problem is an optimization task that tries to answer the question: "What can we do to influence more potential customers to sign up for the company's services?" or "Given our environment and constraints, what should we do to improve our profitability?" The idea is to find the action(s) the company needs to take to migrate potential customers from 'non-join' to 'join'. That is, the objective is not just to predict the behavior of potential customers but to try and influence their behavior in a way that will benefit the company.

Proactive data mining is well positioned to help the security company achieve its objectives. To tackle the problem at hand, we had to take many preparation steps. In this chapter, we will describe several of these steps to demonstrate concept principles.

4.2.2   The Security Dataset

The security company dataset contains 1,600 records of potential customers (both those that independently contacted the company and those the company contacted) and their respective decisions (i.e., to sign up or not for the company's security services) for a given month. Half (50 %) of the customer records in the dataset are labeled as "yes" (accept the company's offer)and 50 % as "no" (reject the company's offer.)

Although the original security company dataset consists of many attributes, for the sake of simplicity, we use only the following most relevant attributes:

(i) Customer-type = {"business", "private"}: Differentiate between two types of customers: private (i.e., homes) and business.
(ii) Customer-size = {"large", "medium", "small"}: An indicator of size category of the different customers. Customers defined as "business" can be categorized as large, medium or small. All private customers are considered "small".
(iii) Contact-initiation = {"customer", "company"}: Indicates who initiated the contact with the customer. This attribute can have one of two values: company initiated the contact or the potential customer called the company.
(iv) Call-or-Visit = {"call", "visit"}: Indicates whether the contact was through a phone call or a physical visit to the customer's site. Obviously, a call is less expensive for the company than an actual visit to the customer's site.

(v) Location = {"center", "center-center", "close-periphery", "far-periphery"}: Indicates the customer's general location. The company divided the country into these four areas based on the locations of its surveillance centers. Each such attribute value has ramifications on the amount of money the company must invest in order to get the customer to sign up. That is, a customer located in the center-center necessitates a short drive from the company's offices while a customer in the far-periphery entails a longer drive and different approach. Obviously, servicing a nearby customer costs less than a customer who lives further away.

(vi) Reference = {"yes", "no"}: Indicates whether this customer was referred to the company or not. Usually the references are made by existing customers of the company.

(vii) Cost-offer = {continues}: The suggested monthly payment by the customer for the company's security services. This amount depends on different parameters (i.e., customer size, location, equipment needed, etc.).

(viii) Sales-Person = {"a", "b", "c"}: Indicates the salesperson assigned by the company to handle the customer. For the sake of simplicity, we divided all sales personnel of the company into three categories based on specific criteria provided by the company (i.e., seniority, level of education, past success rates, etc.).

(ix) Potential-other-Income = {continues}: This attribute is a "silent" attribute (i.e., it is used in calculations during the construction process of the decision tree but does not 'participate' in the tree). This attribute represents potential income the company can earn from the customer should he sign up for the company's services. This income includes income from selling equipment (i.e., sensors, cameras, etc.). The amount of potential other income is a function of the size of the customer and his needs.

(x) Accept? = {"Yes", "No"}: This target attribute indicates the customer's decision to accept or reject the company's offer. It is important to note that this attribute has a benefit matrix that indicates the potential benefit (or loss) to the company should the customer join or not join (i.e., accept or reject) the company's services. This benefit (or loss) is different for each customer based on size, location, equipment, existing referring customer, etc.

4.2.3 Attribute Discretization

The security company's dataset consists of many attributes. The attributes mentioned above were found to be the most relevant to the company's objectives. Since several of the selected attributes had many different values, it was decided to streamline them. The "Customer-Offer" attribute, for example, had over 50 different values. To streamline the data, we arrived at four types of monthly payments representing the payments that the majority of the existing customers made. Also, the "location" attribute was streamlined by dividing the country into four zones (center-center, center, close-periphery and far-periphery) instead of over 200 locations in the original dataset. Similarly, several other attributes were streamlined according to the business objectives.

4.2.4   Additional Environment and Problem Knowledge for the Security Company

As discussed above, a major difference between the proactive and the passive data mining methods is how the proactive method considers specific problems and environmental knowledge when the decision model is being constructed and during the optimization phase. As a result of these processes, the constructed models and the actions suggested by the optimization process are more relevant to the problem at hand.

In addition to the dataset above, there is other knowledge it should consider in the model construction and optimization processes:

(i) From past experience the company knows that the average amount of time a customer will remain with the company is about 3 years. This is due to the initial cost for acquiring the required equipment (i.e., alarm system, cameras, sensors, etc.) which 'ties' the customer to the company.

(ii) From past experience, the company knows that when a potential customer is referred to the company by another customer, the chances of signing the potential customer are much higher.

(iii) It is also known that unless the incentive offered to its existing customer is high (above 3 months of free service), there is very little chance of getting existing clients to refer new potential clients to the company. This information is very important. As can be seen below, we set the incentive at 6 months of free service (see cost matrix for Contact-Initiated attribute.) We arrive at this number after the company performed several tests with a small number of customers in order to determine their degree of responsiveness. Six months of free service was chosen as having the best possibility for encouraging existing customers to provide new referrals while enabling the company to maintain a reasonable level of profitability. Note that this offer was only granted if the referred customer signed with the company.

In addition to the above knowledge, for the proactive method to construct a model that maximizes the utility function set by the organization, it must also know all the costs and benefits associated with any possible change in an attribute. The following describes these costs and benefits for the security company.

4.2.4.1   Cost Matrices

Although the company's objective is to sign up new customers, it must maintain a minimum level of profitability from each customer. Therefore, we have to set the

Table 4.14 Cost matrix for customer-type

	Private	Business
Private	0	∞
Business	∞	0

Table 4.15 Cost matrix for customer-size

	Small	Medium	Large
Small	0	∞	∞
Medium	∞	0	∞
Large	∞	∞	0

Table 4.16 Cost matrix for location

	Center-center	Center	Far-periphery	Close-periphery
Center-center	0	∞	∞	∞
Center	∞	0	∞	∞
Far-periphery	∞	∞	0	∞
Close-periphery	∞	∞	∞	0

Table 4.17 Cost matrix for call-or-visit

	Call	Visit
Call	0	200
Visit	0	0

Table 4.18 Cost matrix for reference

	Yes	No
Yes	0	0
No	(6 * Cost-offer)	0

matrices in a way that reflects the exact costs and benefits associated with a customer. We set the different cost and benefit matrices for all possible attribute values. Their objective is to identify the potential actions that could maximize the number of singed up customers. For that, we have to set the different cost metrics to reflect the company's knowledge and experience.

The following are the cost matrices for all attributes indicated above. We set an infinite cost for such attributes as Customer-type, Customer-size, and Location (as shown in Tables 4.14, 4.15, 4.16) to indicate that any changes to the values of these attributes are prohibited.

The cost matrix for the "Call-or-Visit" (described in Table 4.17) attribute reflects the potential cost of a visit by a sales person at the potential customer's site. The assigned cost reflects the cost to the company of sending a sales person to meet a potential customer. Note that changing from 'Visit' to 'Call' results in 0 refund (i.e., negative cost) since the company can't recover the expense of the visit.

The company's past experience shows that when a customer is referred to the company by another client, the chances of signing the referred individual are enhanced. Therefore, the company sets the cost matrix of the reference attribute as shown in Table 4.18, to reflect this reality. That is, should it offer their existing customers an average of 6 months of free services for each potential customer referred to the company, about 30 % of the customers will accept the offer and refer potential contacts.

Table 4.19 Cost matrix for contact-initiated

	Company	Customer
Company	0	(6 * Cost-offer)
Customer	∞	0

Table 4.20 Cost matrix for cost-offers

Changing to	Cost function
< Current "Cost-offer"	36 * (Current Cost-offer—Target Cost-offer)
> Current "Cost-offer"	∞

The contact-initiated attribute indicates who initiated the contact between the company and the client. Some initiated the contact without being referred by other customers; some are due to referrals; and others were contacted by the company (from purchased lists). We set the cost matrix for "contact-initiated" such that when a customer initiated the contact with the company, it is not possible to change the attribute value (i.e., an infinite cost).However, when the client has no reference, we set the changing cost of the attribute to reflect the cost for finding an existing customer that will be used as referral. That cost indicates the incentive the existing customer receives for his referral. Since we recognize the fact that it is not possible to find a referring customer for each non-referred customer who contacted the company, we multiplied the total number of customers by a confidence factor (provided by the company from past experience). In addition, the idea is to demonstrate to the company that if they initiate a promotional campaign encouraging their current customers to refer new prospects, the system will present the "boundaries" of the incentive the company can offer and within which it is worthwhile for the company to operate, as shown in Table 4.19.

The matrix for the cost-offer attribute reflects company policy to allow reductions but not increases in monthly payments by the customer. When the company considers a reduction in the monthly payment (i.e., cost-offer) of a particular customer (or group of customers), the potential "cost" to the company is the loss of income from that customer (or group of customers) for the average expected time the customer (or group of customers) stays with the company. On the other hand, to prevent the system from considering an increase in monthly payments (since it assumes the customers will resist such changes), it assigns an infinite cost to increases in the value of the cost-offer attribute, as shown in Table 4.20.

The change in the values of the salesperson attribute has no cost. The company does not incur any cost by assigning different salesperson to a given customer.

4.2.4.2   Benefit Matrix

The benefit matrix represents the positive and negative gains associated with each target class value. The dataset we have collected includes an acceptance indication (i.e., subscribe to company's services). Although the problem facing the security company is not a typical prediction, we use the acceptance indication to assist in

Table 4.21 Benefit matrix for the security company

Accept	Benefit
Yes	((36 * Cost-offer) + potential other income)
No	(36* (− Cost-offer—potential other income))

determining the potential accumulated benefit (or loss) expected from clients. When the company considers moving customers from "No" to "Yes" (i.e., not-subscribe to subscribe), it considers the benefit loss (or gain) by reducing the potential new income (new customer) from the old income (existing customer) in addition to the other potential income from each customer (see the "silent attribute", potential-other-income). This benefit is described in Table 4.21. The objective is to find the best actions that are still beneficial despite the change in income.

Note that the benefit (or loss) to the company is calculated at 36 times the monthly income (in addition to other potential income) because the company knows that once signed up a customer will remain with the company an average of at least 3 years.

4.2.5 Passive Classification Model for the Security Company

Given the above dataset, the traditional passive classification methods could help find the characteristics of the subgroups of potential customers that are most likely to accept or reject the security company's offer. This information can be helpful to the company in determining which potential customers are the best candidates to register for its services and which are not. The passive data mining models use only the dataset as input and ignore any additional environmental or problem specific knowledge. The passive model we generated using a J48 implementation of Weka (Waikato Environment for Knowledge Analysis, 1999–2011) to analyze the company's dataset was subsequently compared to our Maximal Utility proactive generated model. Figure 4.4 describes the J48 generated decision model.

4.2.6 Maximal Utility Generated Model for the Security Company

The cost and benefit matrices represent the specific domain knowledge that must be incorporated in the proactive model to become applicable. The decision tree that we produced uses the Maximal Utility splitting criterion along with the cost and benefit functions and the combined dataset of 1600 customer records described above. The output decision tree is presented in Fig. 4.5. The initial overall pessimistic benefit of the tree is: TB(DT) = 384,000. The accuracy of the tree as measured by a splitting criterion with 10-fold cross validation is 73.13 %. As in the J48 model (Fig. 4.4), the "Call or Visit" attribute is the root node in this model. However, the model is more detailed based on the attribute "sales person" since the cost matrix for this attribute is

- Call or Visit = "call": no (800/220)
- Call or Visit = "visit"
  - reference = "no"
    - sales person = "a": no (80/40)
    - sales person = "b"
      - Customer Size = "small": no (0/0)
      - Customer Size = "medium": no (40/0)
      - Customer Size = "large": yes (80/20)
    - sales person = "c": no (40/0)
  - reference = "yes": yes (560/80)

Number of leaves: 7
Number of attributes: 10
Total Benefit: 384000
Tree Accuracy: 72.5
Correctly Classified Rows: 1240
Incorrectly Classified Rows: 360

Fig. 4.4 J48 generated model for the security company

0. Consequently the model "provides" many more options for "moves" (i.e., actions) that the company can implement.

So far we have generated two decision trees. One uses the passive data mining method J48 (Fig. 4.4) while the second is based on our proactive Maximal Utility method (Fig. 4.5). Note that both models have the same initial total benefit of $TB(DT) = 9,313,363$. This reflects the current benefit of all 1,600 customers in the dataset. Since the company's objective is to find the potentially beneficial actions or moves that maximize the utility function set by the company (i.e., more profitable customers), we need to perform the second phase of our proactive method running the optimization algorithm for both decision trees (Figs. 4.2 and 4.3) and comparing the potential utility gains that were generated.

4.2.7 Optimization Algorithm over the J48 Generated Model for the Security Company

To generate the potentially beneficial moves (i.e., actions) from the J48 model, we must perform the second phase of the utility maximization method (i.e., run the optimization algorithm—see Chap. 2)for the J48 decision tree (Fig. 4.4). The optimization algorithm systematically scans the branches of the decision tree, and extracts a list of the advantageous single-branch moves(Matatov et al. 2010). Running the

- Call or Visit = "call"
  - sales person = "a"
    - reference = "no"
      - Customer Size = "medium": no (120/0)
      - Customer Size = "small"
        - cost-offer > 300: no (120/0)
        - cost-offer <= 300: no (120/60)
    - reference = "yes"
      - Customer Type = "private": no (40/20)
  - sales person = "b"
    - reference = "no"
      - Customer Type = "private"
        - location = "center-center"
          - Customer Size = "small": no (40/0)
        - location = "far-periphery": no (120/60)
    - reference = "yes"
      - Customer Type = "business": no (40/20)
      - Customer Type = "private"
        - Contact-Initiation = "Company"
          - Customer Size = "small": no (40/0)
        - Contact-Initiation = "Customer": no (40/20)
  - sales person = "c"
    - cost-offer > 300
      - Customer Type = "business"
        - Customer Size = "medium": no (40/0)
    - cost-offer <= 300
      - Customer Type = "private": no (80/40)
- Call or Visit = "visit"
  - sales person = "a"
    - reference = "no"
      - Customer Size = "large": no (40/20)
      - Customer Size = "medium": no (40/20)
    - reference = "yes": yes (40/0)
  - sales person = "b"
    - reference = "no"
      - Customer Size = "large": yes (80/20)
      - Customer Size = "medium"
        - Customer Type = "business": no (40/0)
    - reference = "yes"
      - Customer Type = "business": yes (200/20)
  - sales person = "c"
    - reference = "no"
      - Customer Type = "business"
        - Customer Size = "large": no (40/0)
    - reference = "yes"
      - Customer Size = "large"
        - Customer Type = "business": yes (120/40)
      - Customer Size = "medium": yes (40/20)
      - Customer Size = "small": yes (160/20)

Number of leaves: 21
Number of attributes: 49
Total Benefit: 384000
Tree Accuracy: 73.13%
Correctly Classified Rows: 1240
Incorrectly Classified Rows: 360

Fig. 4.5 Maximal Utility generated model for the security company

Table 4.22 Optimized algorithm generated action list over the J48 model for the security company

From branch (β_i)	To branch (β_j)	Utility (β_i, β_j)
Call or Visit = "call"	Call or Visit = "visit" AND reference = "no" AND sales person = "a"	1,332,220.83
Call or Visit = "visit" AND reference = "no" AND sales person = "c"	Call or Visit = "visit" AND reference = "no" AND sales person = "b" AND Customer Size = "large"	604,475.66
Call or Visit = "visit" AND reference = "no" AND sales person = "b" AND Customer Size = "medium"	Call or Visit = "visit" AND reference = "no" AND sales person = "a"	219,444.24

optimization algorithm on the J48 decision tree results in the actions, which are described in Table 4.22.

According to the system, "visiting" customers is more productive than "calling" them (suggested action 1). Also, it points out that when there are no "references", salesperson "b" is more productive than "c" (suggested action 2), and salesperson "a" is more productive than "b" for medium-sized customers (suggested action 3).

As we can see, the initial benefit of the model is 384,000. The company should carefully review all the actions the system suggests and decide which one, if any, should be implemented. Should the company choose to implement all of the suggested actions, the additional overall utility gain for the company is: Utility $(\beta_i, \beta_j) = 2,156,140.73$. This represents a total utility gain of 560 % over the initial pessimistic evaluation. As this example indicates, in some cases, even when we perform only the 2nd phase of our proactive algorithm (i.e., optimization) on passive data mining generated models, we still can achieve some reasonable benefits. However, the potential additional benefit is always lower than the one produced by the proactive data mining model.

4.2.8   Optimization Algorithm over the Maximal Utility Generated Model for the Security Company

In contrast to the J48 model, the optimization algorithm running on the maximized utility decision tree (Fig. 4.5) results in the various potential actions presented in Table 4.22. The method considers the knowledge provided during the construction of the decision tree with the objective of determining future interventions that can maximize the utility function the organization sets itself. Comparing both suggested action lists (Tables 4.22 and 4.23), it is clear that given both the same dataset and additional knowledge (i.e., cost and benefit matrices) the decision tree generated by the proactive Maximal Utility method provides far more options for intervention and potentially much higher utility gain than the decision tree generated by the passive J48 technique.

Table 4.23 Optimized algorithm generated action list over the Maximal Utility model for the security company

From branch (β_i)	To branch (β_j)	Utility (β_i, β_j)
Call or Visit = "visit" AND sales person = "c" AND reference = "yes" AND Customer Size = "small"	Call or Visit = "visit" AND sales person = "a" AND reference = "yes"	1,488,000
Call or Visit =, "visit" AND sales person = "b" AND reference = "yes" AND Customer Type = "business"	Call or Visit = "visit" AND sales person = "a" AND reference = "yes"	960,000
Call or Visit = "visit" AND sales person = "c" AND reference = "yes" AND Customer Size = "large" AND Customer Type = "business"	Call or Visit = "visit" AND sales person = "a" AND reference = "yes"	960,000
Call or Visit = "call" AND sales person = "c" AND cost-offer ≤ 300 AND Customer Type = "private"	Call or Visit = "visit" AND sales person = "a" AND reference = "yes"	944,000
Call or Visit = "call" AND sales person = "c" AND cost-offer > 300 AND Customer Type = "business" AND Customer Size = "medium"	Call or Visit = "visit" AND sales person = "a" AND reference = "yes"	832,000
Call or Visit = "call" AND sales person = "a" AND reference = "no" AND Customer Size = "medium"	Call or Visit = "visit" AND sales person = "a" AND reference = "no" AND Customer Size = "medium"	683,347.73
Call or Visit = "call" AND sales person = "b" AND reference = "yes" AND Customer Type = "private" AND Contact-Initiation = "Company" AND Customer Size = "small"	Call or Visit = "visit" AND sales person = "a" AND reference = "yes"	616,000
Call or Visit = "visit" AND sales person = "c" AND reference = "no" AND Customer Type = "business" AND Customer Size = "large"	Call or Visit = "visit" AND sales person = "b" AND reference = "no" AND Customer Size = "large"	604,475.67
Call or Visit = "call" AND sales person = "b" AND reference = "yes" AND Customer Type = "private" AND Contact-Initiation = "Customer"	Call or Visit = "visit" AND sales person = "a" AND reference = "yes"	472,000
Call or Visit = "visit" AND sales person = "b" AND reference = "no" AND Customer Size = "medium" AND Customer Type = "business"	Call or Visit = "visit" AND sales person = "a" AND reference = "no" AND Customer Size = "medium"	235,782.58
Call or Visit = "visit" AND sales person = "a" AND reference = "no" AND Customer Size = "large"	Call or Visit = "visit" AND sales person = "b" AND reference = "no" AND Customer Size = "large"	201,491.89
Call or Visit = "call" AND sales person = "a" AND reference = "no" AND Customer Size = "small" AND cost-offer > 300	Call or Visit = "call" AND sales person = "a" AND reference = "no" AND Customer Size = "small" AND cost-offer ≤ 300	136,412.17
Call or Visit = "visit" AND sales person = "c" AND reference = "yes" AND Customer Size = "medium"	Call or Visit = "visit" AND sales person = "a" AND reference = "yes"	120,000
Call or Visit = "call" AND sales person = "b" AND reference = "no" AND Customer Type = "private" AND location = "center-center" AND Customer Size = "small"	Call or Visit = "call" AND sales person = "c" AND cost-offer ≤ 300 AND Customer Type = "private"	87,777.70

It should be noted that when all domain knowledge provided by the company is considered, the first five suggested actions by the system simply indicate that salesperson "a" is more successful in convincing customers to join the company's services than the other sales people. The sixth suggested action is for salesperson "a" to improve his results by making personal "visits" to medium-sized customers with "no" references rather than "call". Action seventh suggested action is similar to the second suggested action for "small" "private" customers who initiated the contact. Suggested action eight implies that for "large" customers with "no" references, the company should send sales person "b" rather than "c" to "visit" customers. Action nine suggests that for "private" customers, salesperson "a" should make "visits" rather than "call".

As in the J48 model (Fig. 4.4), the initial benefit of the Maximal Utility model (Fig. 4.5) is also 384,000. The company should carefully review the actions suggested by the system and decide which, if any, should be implemented. Should the company choose to implement all of above suggested actions, the additional overall utility gain for the company is: Utility $(\beta_i, \beta_j) = 8,341,302.74$. This represents a total utility gain of 2,172 % over the initial pessimistic benefit. Obviously, it is unlikely that the company will implement all suggested actions. However it provides the organization with many options to choose from. It is important to note that there are no guarantees that these actions will necessarily work. However, it focuses the company towards the areas it need to look at in order to potentially maximize its benefits. Clearly, the proactive data mining model provides many more potential beneficial actions and far greater potential utility gains than the models generated by traditional passive inducers like the J48.

4.3 Case Studies Summary

This chapter described real-life scenarios facing two leading Israeli companies. Using the proactive classification approach, we demonstrated that by using problem specific and environment knowledge, in the form of attribute-change and benefit functions the typical classification task can be expanded to support a very real critical business need of the companies—business optimization.

Although the descriptions are based on real companies and their data, these are not closed-loop case studies because various validation aspects are lacking. We have no data regarding the actual execution of the recommended actions or the actual utility that was gained from that execution. The recommended actions are incomplete but even if they were not, we would still not recommend them as inputs for automatic implementation. Instead, we suggest our recommendations be regarded as input for human evaluation. Accordingly, the result actions can easily be interpreted by humans.

The above examples demonstrate the potential applications of our proactive data mining approach. The main concept is based on the assumption that a system can operate more intelligently and with a greater degree of relevance to real world problems if it knows more about the problem it is trying to solve. This assumption reflects the knowledge principle defined by (Feigenbaum 1989) which states: "a system exhibits intelligent understanding and action at a high level of competence primarily because of the specific knowledge that it contains about its domain of endeavor". According to Feigenbaum (1989), the main conclusion from this principle is that reasoning processes of an intelligent system, being general and therefore weak, are not the source of power that leads to high levels of competence in behavior. The knowledge principle simply says that if a program is to perform well, it must know a great deal about the world in which it operates. In the absence of that specific domain knowledge, reasoning won't help.

References

Arbel R, Rokach L (2006) Classifier evaluation under limited resources. Pattern Recogn Lett 27(14):1619–1631

Feigenbaum E A (1989) Toward the library of the future. Long Range Plann 22(1):118–123

Hung S-Y, Yen DC, Wang H-Y (2006) Applying data mining to telecom churn management. Expert Syst Appl 31:515–524

Matatov N, Rokach L, Maimon O (2010) Privacy-preserving data mining: a feature set partitioning approach. Inform Sci 180(14):2696–2720

Menahem E, Rokach L, Elovici Y (2009) Troika—an improved stacking schema for classification tasks. Inform Sci 179(24):4097–4122

Rokach L, Maimon O, Lavi I (2003) Space decomposition in data mining: a clustering approach. Found Intell Syst 24–31

Chapter 5
Sensitivity Analysis of Proactive Data Mining

As stated in Chap. 4, to achieve an effective and applicable solution for a data mining problem, it is vital to thoroughly understand the problem at hand, in particular its constraints, environment and its problem specific knowledge. However, it is difficult to pinpoint the exact knowledge (i.e., attribute values) necessary for optimally implementing the proactive data mining method. In this chapter we present several scenarios over the security company's case (Chap. 4) to demonstrate the general boundaries of the method.

5.1 Zero-one Benefit Function

The zero-one loss scenario is the basic operating mode of most existing passive classification systems (such as J48, CART, etc.). Countless research papers have been published based on comparisons of classifier accuracy over benchmark datasets (Breiman et al. 1996; Fayyad and Irani 1992; Buntine and Niblett 1992; Loh and Shih 1997, 1999; Provost and Fawcett 1997; Lim et al. 2000). On the other hand, Provost and Fawcett (1998) argue that comparing accuracies on benchmark datasets say little, if anything, about classifier performance on real world tasks. The reason is that most research in machine learning considers all misclassification errors as having an equivalent cost. In real world problems, this is rarely the case (Rokach 2006, 2009). However, in the scenario below, we operate the proactive data mining method with a zero-one benefit function and with all cost matrix values of zero (i.e., for all entries, all attributes, and for any attribute value change), as shown in Table 5.1. We then compare the results with those of a passive zero-one loss method (i.e., J48):

When neither domain knowledge nor constraints are considered, the proactive method offers no major benefits compared to the passive approaches (i.e., J48). The two models below were generated for the security company's dataset by J48 (Fig. 5.1) and by the proposed Maximal Utility algorithms (Fig. 5.2)

As we can see from the models, the "sales person" and "reference" attributes are dominant in both models. Even the accuracy of the two models is similar. However, the proactive model is more detailed. The detailed model allows for more potential actions during the optimization phase of the proactive data mining method. As noted

H. Dahan et al., *Proactive Data Mining with Decision Trees*,
SpringerBriefs in Electrical and Computer Engineering,
DOI 10.1007/978-1-4939-0539-3_5, © The Author(s) 2014

Table 5.1 Zero-one benefit for the security company

Accept	Benefit
Yes	1
No	0

above, when the question at hand is a simple zero-one loss, then the Maximal Utility method is an additional classification method offering a reasonable degree of accuracy. Although zero-one loss has very small relevance to real world problems since it assumes the same 'cost' for all customers, we executed the optimization phase of the proactive data mining algorithm on both models to see what potential actions would emerge. Obviously, since there are no costs associated with any attribute value changes, the optimization algorithm may suggest various unreasonable actions such as a change of customer location or other similar unalterable attributes. Such pitfalls are another important indication of the critical need to consider environment and problem specific knowledge for solution applicability (Kisilevich et al. 2010). Table 5.2 lists all the potential actions that the optimization algorithm suggests over the J48 generated model (Fig. 5.1):

Table 5.3 lists actions suggested by the optimization algorithm over the Maximal Utility model with zero-one benefit function (Fig. 5.2).

Given the fact that there are no associated costs for any attribute value changes and with zero-one benefit function, the path classifying the highest number of positive instances will be selected as the target path for all actions. In this case, the utility

- Call or Visit = "call": no (800/220)
- Call or Visit = "visit"
  - reference = "no"
    - sales person = "a": no (80/40)
    - sales person = "b"
      - Customer Size = "small": no (0/0)
      - Customer Size = "medium": no (40/0)
      - Customer Size = "large": yes (80/20)
    - sales person = "c": no (40/0)
  - reference = "yes": yes (560/80)

Number of leaves: 7
Number of attributes: 10
Total Benefit: 800
Tree Accuracy: 75.81%
Correctly Classified Instances: 1213
Incorrectly Classified Instances: 387

Fig. 5.1 J48 generated model for the security company

- location = "center"
  - reference = "no"
    - sales person = "a"
      - Customer Type = "business"
        - Customer Size = "medium": no (40/20)
    - sales person = "b"
      - Customer Type = "business"
        - Customer Size = "medium"
          - Contact-Initiation = "Customer": no (40/0)
  - reference = "yes"
    - sales person = "b": yes (200/20)
    - sales person = "c": yes (40/0)
- location = "center-center"
  - cost-offer > 300
    - sales person = "a"
      - Customer Size = "medium"
        - Customer Type = "business"
          - Contact-Initiation = "Company": no (120/0)
      - Customer Size = "small"
        - Customer Type = "business"
          - Contact-Initiation = "Company": no (120/0)
    - sales person = "b"
      - Customer Type = "business"
        - Customer Size = "small": no (40/20)
    - sales person = "c"
      - Customer Type = "business"
        - Customer Size = "medium"
          - Contact-Initiation = "Company": no (40/0)
  - cost-offer <= 300
    - sales person = "a"
      - Customer Type = "private"
        - Customer Size = "small": no (160/80)
    - sales person = "b"
      - Customer Type = "private"
        - Customer Size = "small"
          - Contact-Initiation = "Company": no (40/0)
          - Contact-Initiation = "Customer": no (40/0)
    - sales person = "c"
      - Customer Type = "private"
        - Customer Size = "small": no (80/40)
- location = "close-periphery"
  - sales person = "a"
    - reference = "no"
      - Customer Type = "business"
        - Customer Size = "large": no (40/20)
    - reference = "yes": yes (40/0)
  - sales person = "b"
    - Customer Type = "business"
      - Customer Size = "large": yes (80/20)

Fig. 5.2 Maximal utility generated model for the security company with zero cost matrices and zero-one benefit function

- sales person = "c"
  - reference = "no"
    - Customer Type = "business"
      - Customer Size = "large"
        - Contact-Initiation = "Company": no (40/0)
  - reference = "yes"
    - Customer Type = "business"
      - Customer Size = "large": yes (120/40)
- location = "far-periphery"
  - Call or Visit = "call"
    - Customer Type = "private"
      - Customer Size = "small"
        - Contact-Initiation = "Customer"
          - reference = "no": no (120/60)
          - reference = "yes": no (40/20)
  - Call or Visit = "visit"
    - Customer Type = "private"
      - Customer Size = "small": yes (160/20)

Number of leaves: 20
Number of attributes: 67
Total Benefit: 800
Tree Accuracy: 76.37%
Correctly Classified Instances: 1240
Incorrectly Classified Instances: 360

Fig. 5.2 (continued)

Table 5.2 Optimized algorithm generated action list over the J48 model of the security company with zero-one benefit function

From Branch (β_i)	To Branch (β_j)	utility (β_i, β_j)
Call or Visit = "call"	Call or Visit = "visit" AND reference = "yes"	385.04
Call or Visit = "visit" AND reference = "no" AND sales person = "c"	Call or Visit = "visit" AND reference = "yes"	28.35
Call or Visit = "visit" AND reference = "no" AND sales person = "b" AND Customer Size = "medium"	Call or Visit = "visit" AND reference = "yes"	28.35
Call or Visit = "visit" AND reference = "no" AND sales person = "a"	Call or Visit = "visit" AND reference = "yes"	23.62
Call or Visit = "visit" AND reference = "no" AND sales person = "b" AND Customer Size = "large"	Call or Visit = "visit" AND reference = "yes"	7.09

Table 5.3 Optimized algorithm generated action list over the Maximal Utility model for the security company with zero-cost matrices and zero-one benefit function

From Branch (β_i)	To Branch (β_j)	utility (β_i, β_j)
Location = "center-center" AND cost-offer > 300 AND sales person = "a" AND Customer Size = "small" AND Customer Type = "business" AND Contact-Initiation = "Company"	Location = "close-periphery" AND sales person = "a" AND reference = "yes"	111.47
Location = "center-center" AND cost-offer > 300 AND sales person = "a" AND Customer Size = "medium" AND Customer Type = "business" AND Contact-Initiation = "Company"	Location = "close-periphery" AND sales person = "a" AND reference = "yes"	111.47
Location = "center-center" AND cost-offer ≤ 300 AND sales person = "a" AND Customer Type = "private" AND Customer Size = "small"	Location = "close-periphery" AND sales person = "a" AND reference = "yes"	74.32
Location = "far-periphery" AND Call or Visit = "call" AND Customer Type = "private" AND Customer Size = "small" AND Contact-Initiation = "Customer" AND reference = "no"	Location = "close-periphery" AND sales person = "a" AND reference = "yes"	55.74
Location = "close-periphery" AND sales person = "c" AND reference = "yes" AND Customer Type = "business" AND Customer Size = "large"	Location = "close-periphery" AND sales person = "a" AND reference = "yes"	37.16
Location = "close-periphery" AND sales person = "c" AND reference = "no" AND Customer Type = "business" AND Customer Size = "large" AND Contact-Initiation = "Company"	location = "close-periphery" AND sales person = "a" AND reference = "yes"	37.16
Location = "center-center" AND cost-offer ≤ 300 AND sales person = "c" AND Customer Type = "private" AND Customer Size = "small"	Location = "close-periphery" AND sales person = "a" AND reference = "yes"	37.16
Location = "center-center" AND cost-offer ≤ 300 AND sales person = "b" AND Customer Type = "private" AND Customer Size = "small"	Location = "close-periphery" AND sales person = "a" AND reference = "yes"	37.16
Location = "center-center" AND cost-offer ≤ 300 AND sales person = "b" AND Customer Type = "private" AND Contact-Initiation = "Customer"	Location = "close-periphery" AND sales person = "a" AND reference = "yes"	37.16
Location = "center-center" AND cost-offer > 300 AND sales person = "b" AND Customer Type = "private" AND Contact-Initiation = "Company"	Location = "close-periphery" AND sales person = "a" AND reference = "yes"	37.16
Location = "center-center" AND cost-offer > 300 AND sales person = "c" AND Customer Type = "business" AND Customer Size = "medium" AND Contact-Initiation = "Company"	Location = "close-periphery" AND sales person = "a" AND reference = "yes"	37.16
Location = "center" AND reference = "no" AND sales person = "b" AND Customer Type = "business" AND Customer Size = "medium" AND Contact-Initiation = "Customer"	Location = "close-periphery" AND sales person = "a" AND reference = "yes"	37.16
Location = "far-periphery" AND Call or Visit = "visit" AND Customer Type = "private" AND Customer Size = "small"	Location = "close-periphery" AND sales person = "a" AND reference = "yes"	18.58
Location = "far-periphery" AND Call or Visit = "call" AND Customer Type = "private" AND Customer Size = "small" AND Contact-Initiation = "Customer" AND reference = "yes"	Location = "close-periphery" AND sales person = "a" AND reference = "yes"	18.58

Table 5.3 (continued)

From Branch (β_i)	To Branch (β_j)	utility (β_i, β_j)
Location = "close-periphery" AND sales person = "b" AND Customer Type = "business" AND Customer Size = "large"	Location = "close-periphery" AND sales person = "a" AND reference = "yes"	18.58
Location = "close-periphery" AND sales person = "a" AND reference = "no" AND Customer Type = "business" AND Customer Size = "large"	Location = "close-periphery" AND sales person = "a" AND reference = "yes"	18.58
Location = "center-center" AND cost-offer > 300 AND sales person == "b" AND Customer Type = "business" AND Customer Size = "small"	Location = "close-periphery" AND sales person = "a" AND reference = "yes"	18.58
Location = "center" AND reference = "no" AND sales person = "a" AND Customer Type = "business" AND Customer Size = "medium"	Location = "close-periphery" AND sales person = "a" AND reference = "yes"	18.58
Location = "center" AND reference = "yes" AND sales person = "b"	Location = "close-periphery" AND sales person = "a" AND reference = "yes"	18.58

Table 5.4 Dynamic benefits function for the security company

Accept	Benefit
Yes	(Cost-offer)
No	(-Cost-offer)

function (see Chap. 3) reflects simply the difference between the benefits of the selected target (β_i) and the source (β_j) paths multiplied by the weight. That is:

$$\text{utility}\left(\beta_j, \beta_i\right) = \left[TB\left(\beta_i\right) \cdot \frac{\left|v\left(|\beta_j|\right)\left(\langle X; Y\rangle\right)\right|}{\left|v\left(|\beta_{2i}|\right)\left(\langle X; Y\rangle\right)\right|} - TB\left(\beta_j\right) \right] \cdot w\left(\beta_j, \beta_i\right) - 0$$

The optimization algorithm selects a target path that maximizes the benefit function and suggests moving all other paths to the same selected target path. The reason that the target path is the same for all source paths (i.e., all actions) is that the benefit function is fixed (zero-one). Therefore, the path that is selected for the first potential action will be the path with the highest utility (i.e., the path that classifies the most positive instances). Therefore it will also be the highest for all other potential actions. The total potential utility gain from the J48 model is:

$$\text{utility}\left(\beta_i, \beta_j\right) = 472.45,$$

while the total potential utility gain from the Maximal.

Utility model with zero-one benefit function is: utility $(\beta_i, \beta_j) = 743.16$. It is clear from the results that even with the zero-one lost benefit function, the additional overall potential utility gain from the model is almost double that of the J48 model.

5.2 Dynamic Benefit Function

In this scenario, we assume that all cost matrices are 0 (i.e., for all entries, all attributes, and for any value change to any other value change). However, instead of a zero-one benefit function, the dynamic benefit reflects the potential income expected from each customer. In the security case, it is the monthly payment expected from the customer. This amount is provided to us by the cost-offer attribute. Therefore we can assign that specific value in the benefit matrix as shown in Table 5.4.

As indicated above, when domain knowledge or constraints are not being considered, there is no major benefit to the proactive method compared to passive methods (i.e., J48). Figure 5.3 presents the model generated for the security company's dataset by J48 with the dynamic benefit function (Table 5.4). Obviously, the passive J48 algorithm is indifferent to any benefit function. The only difference between the J48 models in Figs. 5.3 and 5.4 is the total benefit calculated by our optimization algorithm over the J48 model.

Note that the cost-offer is specific to each customer instance. This feature is inherent to the Maximal Utility algorithm. A dynamic benefit function does not exist

- Call or Visit = "call": no (800/220)
- Call or Visit = "visit"
    - reference = "no"
        - sales person = "a": no (80/40)
        - sales person = "b"
            - Customer Size = "small": no (0/0)
            - Customer Size = "medium": no (40/0)
            - Customer Size = "large": yes (80/20)
        - sales person = "c": no (40/0)
    - reference = "yes": yes (560/80)

Number of leaves: 7
Number of attributes: 10
Total Benefit: 32000
Tree Accuracy: 75.81%
Correctly Classified Instances: 1213
Incorrectly Classified Instances: 387

Fig. 5.3 J48 model for the security company (with dynamic benefit)

in most traditional classification methods (i.e., J48). Figure 5.4 presents the Maximal Utility generated model with the dynamic benefit function (Table 5.4):

As we can see, the model is very similar to the Maximal Utility model for the zero-one benefit function (Fig. 5.2). Given the fact that there are no associated costs for any change in attribute value, the optimization algorithm looks for the path that maximizes the utility function with no consideration of costs. The only difference from the zero-one case (Sect. 5.1) is the selected dominant target path (see Table 5.5). In the dynamic benefit function case, the optimization algorithm selects as a target the path that classifies the highly paying customers. The optimization algorithm seeks to maximize the set utility function by "changing" the "behavior" (i.e., attributes or characteristics) of future customers to those of the highly paying customers. Table 5.5 lists the suggested action by the optimization algorithm over the Maximal Utility model (Fig. 5.4) with dynamic benefit.

The total potential utility gain of the Maximal Utility model with the dynamic benefit function is: utility $(\beta_i, \beta_j) = 1,456,584.31$. Given the fact that there are no constraints on the utility gain, the value is extremely high. It is clear from the above results that in both cases, zero-one and dynamic benefit, the additional overall potential utility gain from the model is much higher than from the J48 model. However, these results are probably inapplicable given the unreasonable assumption that there is no associated change cost for any attribute value.

On the other hand, performing the optimization algorithm with the dynamic benefit function and zero change cost for all attributes on the J48 model (Fig. 5.3) will produce the suggested action list, which is shown in Table 5.6.

- location = "center"
  - reference = "no"
    - sales person = "a"
      - Customer Type = "business"
        - Customer Size = "medium"
          - Contact-Initiation = "Customer": no (40/20)
    - sales person = "b"
      - Customer Type = "business"
        - Customer Size = "medium"
          - Contact-Initiation = "Customer"
            - Call or Visit = "visit": no (40/0)
  - reference = "yes"
    - sales person = "b": yes (200/20)
    - sales person = "c": yes (40/0)
- location = "center-center"
  - Customer Type = "business"
    - sales person = "a"
      - Call or Visit = "call"
        - Customer Size = "medium"
          - Contact-Initiation = "Company"
            - reference = "no": no (120/0)
        - Customer Size = "small"
          - Contact-Initiation = "Company"
            - reference = "no": no (120/0)
    - sales person = "b"
      - Customer Size = "small"
        - Contact-Initiation = "Company"
          - Call or Visit = "call": no (40/20)
    - sales person = "c"
      - reference = "yes"
        - Customer Size = "medium"
          - Contact-Initiation = "Company"
            - Call or Visit = "call": no (40/0)

Fig. 5.4 Maximal Utility model for the security company with zero-cost matrices and dynamic benefit function

As we can see from the above list, the optimization algorithm finds the possible potential actions on the J48 models. The total potential utility gain from the J48 model with the dynamic benefit function and zero cost for all attribute value changes is: utility $(\beta_i, \beta_j) = 628,184.62$.

5.3 Dynamic Benefits and Infinite Costs of the Unchangeable Attributes

This scenario is similar to the previous scenario except for the infinite change cost of the unchangeable attributes. That is, in addition to the previous dynamic benefit matrix (Table 5.4), we set to infinite costs the cost matrices for the following

- Customer Type = "private"
  - sales person = "a"
    - reference = "no"
      - Customer Size = "small"
        - Contact-Initiation = "Customer": no (120/60)
    - reference = "yes"
      - Customer Size = "small"
        - Contact-Initiation = "Company": no (40/20)
  - sales person = "b"
    - Customer Size = "small"
      - cost-offer <= 1000
        - Contact-Initiation = "Company"
          - Call or Visit = "call": no (40/0)
        - Contact-Initiation = "Customer"
          - Call or Visit = "call": no (40/0)
  - sales person = "c"
    - Customer Size = "small"
      - Contact-Initiation = "Company"
        - Call or Visit = "call": no (80/40)
- location = "close-periphery"
  - sales person = "a"
    - reference = "no"
      - Customer Type = "business"
        - Customer Size = "large"
          - Contact-Initiation = "Company": no (40/20)
    - reference = "yes": yes (40/0)
  - sales person = "b"
    - Customer Type = "business"
      - Customer Size = "large": yes (80/20)
  - sales person = "c"
    - reference = "no"
      - Customer Type = "business"
        - Contact-Initiation = "Company"
          - Customer Size = "large"
            - Call or Visit = "visit": no (40/0)
    - reference = "yes"
      - Customer Type = "business"
        - Customer Size = "large": yes (120/40)

Fig. 5.4 (continued)

unchangeable attributes: Customer-type, Customer-size, and Location. This directs
the proactive data mining method to consider these unchangeable attributes. This
scenario is closer to reality than the previous no-cost scenario. Figure 5.5 presents
the Maximal Utility model with the dynamic benefit function (Table 5.4) and the
infinite cost for the unchangeable attributes:

- location = "far-periphery"
  - Call or Visit = "call"
    - Customer Type = "private"
      - Customer Size = "small"
        - Contact-Initiation = "Customer"
          - reference = "no": no (120/60)
          - reference = "yes": no (40/20)
    - Call or Visit = "visit"
      - Customer Type = "private"
        - Customer Size = "small"
          - Contact-Initiation = "Customer"
            - reference = "yes": yes (160/20)

Number of leaves: 21
Number of attributes: 83
Total Benefit: 32000
Tree Accuracy: 72.12%
Correctly Classified Instances: 1240
Incorrectly Classified Instances: 360

Fig. 5.4 (continued)

The Maximal Utility model in (Fig. 5.5) is much different than the previous two models (i.e., zero cost matrices with dynamic benefit function). In this model, the Maximal Utility algorithm places the changeable attributes at the top part of the tree (i.e., Call-or-Visit, reference, salesperson) and the unchangeable attributes at the bottom. However, since the change costs of the changeable attributes are zero, this scenario, from the optimization process perspective is very similar, to the previous two scenarios. That is, there is a single dominant target path found by the optimization algorithm that maximizes the utility function. Multiple target paths for different source paths are possible when there are different costs associated with the changeable attributes. In our case, where there are no associated costs with the changeable attributes, the path that maximizes the utility function for the first source path will be the same for all other source paths. Table 5.7 lists potential actions as suggested by the optimization algorithm over the Maximal Utility model (Fig. 5.5) with dynamic benefit function and infinite change cost for the unchangeable attributes.

Unlike the previous scenario, where the single dominant target path that was selected included unchangeable attributes (i.e., location), the selected single dominant target path in this case consists of changeable attributes only (i.e., call or visit, reference and salesperson). The total potential utility gain from the model

Table 5.5 Optimized algorithm generated action list over the Maximal Utility model for the security company with dynamic benefit function

From Branch (β_i)	To Branch (β_j)	utility (β_i, β_j)
Location = "center-center" AND Customer Type = "business" AND sales person = "a" AND Call or Visit = "call" AND Customer Size = "medium" AND Contact-Initiation = "Company" AND reference = "no"	location = "close-periphery" AND sales person = "a" AND reference = "yes"	195078.26
Location = "center-center" AND Customer Type = "business" AND sales person = "a" AND Call or Visit = "call" AND Customer Size = "small" AND Contact-Initiation = "Company" AND reference = "no"	Location = "close-periphery" AND sales person = "a" AND reference = "yes"	167209.93
Location = "far-periphery" AND Call or Visit = "visit" AND Customer Type = "private" AND Customer Size = "small" AND Contact-Initiation = "Customer" AND reference = "yes"	Location = "close-periphery" AND sales person = "a" AND reference = "yes"	115189.06
Location = "far-periphery" AND Call or Visit = "call" AND Customer Type = "private" AND Customer Size = "small" AND Contact-Initiation = "Customer" AND reference = "no"	Location = "close-periphery" AND sales person = "a" AND reference = "yes"	111473.29
Location = "center-center" AND Customer Type = "private" AND sales person = "a" AND reference = "no" AND Customer Size = "small" AND Contact-Initiation = "Customer"	Location = "close-periphery" AND sales person = "a" AND reference = "yes"	111473.29
Location = "close-periphery" AND sales person = "c" AND reference = "yes" AND Customer Type = "business" AND Customer Size = "large"	Location = "close-periphery" AND sales person = "a" AND reference = "yes"	74315.53
Location = "close-periphery" AND sales person == "c" AND reference == "no" AND Customer Type = "business" AND Contact-Initiation == "Company" AND Customer Size = "large" AND Call or Visit = "visit"	Location = "close-periphery" AND sales person = "a" AND reference = "yes"	74315.53
Location = "center-center" AND Customer Type = "private" AND sales person = "c" AND Customer Size = "small" AND Contact-Initiation = "Company" AND Call or Visit = "call"	Location = "close-periphery" AND sales person = "a" AND reference = "yes"	74315.53
Location = "center" AND reference = "yes" AND sales person = "b"	Location = "close-periphery" AND sales person = "a" AND reference = "yes"	74315.53
Location = "center-center" AND Customer Type = "business" AND sales person = "c" AND reference = "yes" AND Customer Size = "medium" AND Contact-Initiation = "Company" AND Call or Visit = "call"	Location = "close-periphery" AND sales person = "a" AND reference = "yes"	65026.08
Location = "center" AND reference = "no" AND sales person = "b" AND Customer Type = "business" AND Customer Size = "medium" AND Contact-Initiation = "Customer" AND Call or Visit = "visit"	Location = "close-periphery" AND sales person = "a" AND reference = "yes"	65026.08

Table 5.5 (continued)

From Branch (β_i)	To Branch (β_j)	utility(β_i, β_j)
Location = "center-center" AND Customer Type = "private" AND sales person = "b" AND Customer Size = "small" AND cost-offer ≤ 1000 AND Contact-Initiation = "Customer" AND Call or Visit = "call"	Location = "close-periphery" AND sales person = "a" AND reference = "yes"	48305.09
Location = "center-center" AND Customer Type = "private" AND sales person = "b" AND Customer Size = "small" AND cost-offer ≤ 1000 AND Contact-Initiation = "Company" AND Call or Visit = "call"	Location = "close-periphery" AND sales person = "a" AND reference = "yes"	48305.09
Location = "far-periphery" AND Call or Visit = "call" AND Customer Type = "private" AND Customer Size = "small" AND Contact-Initiation = "Customer" AND reference = "yes"	Location = "close-periphery" AND sales person = "a" AND reference = "yes"	37157.76
Location = "close-periphery" AND sales person = "b" AND Customer Type = "business"	Location = "close-periphery" AND sales person = "a" AND reference = "yes"	37157.76
Location = "close-periphery" AND sales person = "a" AND reference = "no" AND Customer Type = "business" AND Customer Size = "large"	Location = "close-periphery" AND sales person = "a" AND reference = "yes"	37157.76
Location = "center-center" AND Customer Type = "private" AND sales person = "a" AND reference = "yes" AND Customer Size = "small" AND Contact-Initiation = "Company"	Location = "close-periphery" AND sales person = "a" AND reference = "yes"	37157.76
Location = "center-center" AND Customer Type = "business" AND sales person = "b" AND Contact-Initiation = "Company" AND Call or Visit = "call" AND Customer Size = "small"	Location = "close-periphery" AND sales person = "a" AND reference = "yes"	37157.76
Location = "center" AND reference = "no" AND sales person = "a" AND Customer Type = "business" AND Customer Size = "medium" AND Contact-Initiation = "Customer"	Location = "close-periphery" AND sales person = "a" AND reference = "yes"	37157.76
Location = "center" AND reference = "yes" AND sales person = "c"	Location = "close-periphery" AND sales person = "a" AND reference = "yes"	9289.44

Table 5.6 Optimized algorithm generated action list over the J48 model for the security company with dynamic benefit function

From Branch (β_i)	To Branch (β_j)	utility (β_i, β_j)
Call or Visit = "call"	Call or Visit = "visit" AND reference = "yes"	482840.74
Call or Visit = "visit" AND reference = "no" AND sales person = "c"	Call or Visit = "visit" AND reference = "yes"	48780.14
Call or Visit = "visit" AND reference = "no" AND sales person = "b" AND Customer Size = "medium"	Call or Visit = "visit" AND reference = "yes"	40512.32
Call or Visit = "visit" AND reference = "no" AND sales person = "a"	Call or Visit = "visit" AND reference = "yes"	31417.72
Call or Visit = "visit" AND reference = "yes"	Call or Visit = "visit" AND reference = "no" AND sales person = "b" AND Customer Size = "large"	24633.70

with the dynamic benefit function and infinite cost for the unchangeable attributes is: utility $(\beta_i, \beta_j) = 1,456,584.31$. This potential utility gain (which is identical to one in the previous scenario) is remarkable considering the unchangeable attributes constraint. This scenario demonstrates the way the model considers the problem specific knowledge (i.e., infinite) while constructing the decision model that maximizes the set utility function. On the other hand, running the optimization algorithm with the dynamic benefit function and the unchangeable attributes constraint on the J48 model (Fig. 5.3) will produce the suggested action list, which is shown in Table 5.8.

As we can see from the above list, the optimization algorithm finds the possible actions that include only changeable attributes (Maimon and Rokach 2001). The total potential utility gain from the J48 model with the dynamic benefit function and infinite change cost for the unchangeable attributes is: utility $(\beta_i, \beta_j) = 603,550.92$. This potential utility gain is lower than the one achieved in the previous scenario (Table 5.6) due to the unchangeable attributes constraint.

5.4 Dynamic Benefit and Balanced Cost Functions

This scenario is similar to the previous one except for the different cost for the changeable attribute values (in addition to the infinite change cost for the unchangeable attributes.) The idea is to provide attribute change costs that are close to the benefit. The average monthly payment of the security dataset (i.e., of the cost-offer attribute) is 595. Therefore, to balance the costs, we have to set them below the average to make it worthwhile to suggest possibly actions. If we were to set the costs higher than the benefit, the optimization algorithm would not find any potential actions that would beneficial. Tables 5.9–5.12 provides the cost matrices for the changeable attributes:

This scenario is a good approximation of real world problems where well-run organizations wish to improve their business. Figure 5.6 presents the Maximal Utility

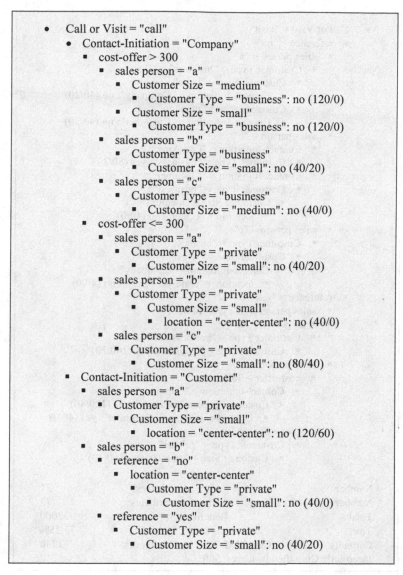

- Call or Visit = "call"
    - Contact-Initiation = "Company"
        - cost-offer > 300
            - sales person = "a"
                - Customer Size = "medium"
                    - Customer Type = "business": no (120/0)
                - Customer Size = "small"
                    - Customer Type = "business": no (120/0)
            - sales person = "b"
                - Customer Type = "business"
                    - Customer Size = "small": no (40/20)
            - sales person = "c"
                - Customer Type = "business"
                    - Customer Size = "medium": no (40/0)
        - cost-offer <= 300
            - sales person = "a"
                - Customer Type = "private"
                    - Customer Size = "small": no (40/20)
            - sales person = "b"
                - Customer Type = "private"
                    - Customer Size = "small"
                        - location = "center-center": no (40/0)
            - sales person = "c"
                - Customer Type = "private"
                    - Customer Size = "small": no (80/40)
    - Contact-Initiation = "Customer"
        - sales person = "a"
            - Customer Type = "private"
                - Customer Size = "small"
                    - location = "center-center": no (120/60)
        - sales person = "b"
            - reference = "no"
                - location = "center-center"
                    - Customer Type = "private"
                        - Customer Size = "small": no (40/0)
            - reference = "yes"
                - Customer Type = "private"
                    - Customer Size = "small": no (40/20)

Fig. 5.5 Maximal Utility model for the security company with infinite change cost for the unchangeable attributes and dynamic benefit function

model with dynamic benefit function (Table 5.4), infinite cost for the unchangeable attributes, and balanced cost for the changeable attributes (Tables 5.13 and 5.14). Figure 5.6 presents a model that is very different from the generated models in the previous scenarios. The Maximal Utility model is different for various cost and benefit values (Ben-Shimon et al. 2007). As indicated above, the passive data mining methods generate decision models regardless of the problem it is solving and with no consideration of any domain or problem specific knowledge. The Maximal Utility algorithm generates models that are adjusted to the specific problem it addresses based

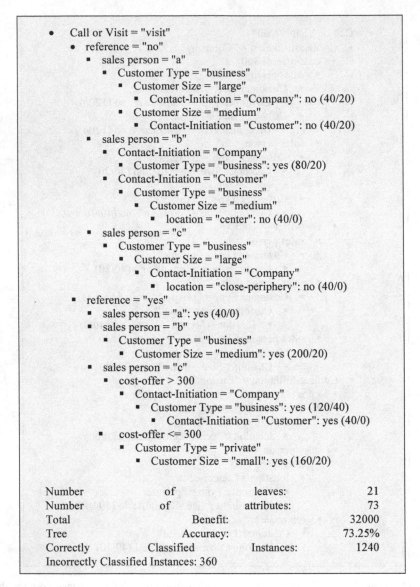

- Call or Visit = "visit"
  - reference = "no"
    - sales person = "a"
      - Customer Type = "business"
        - Customer Size = "large"
          - Contact-Initiation = "Company": no (40/20)
        - Customer Size = "medium"
          - Contact-Initiation = "Customer": no (40/20)
    - sales person = "b"
      - Contact-Initiation = "Company"
        - Customer Type = "business": yes (80/20)
      - Contact-Initiation = "Customer"
        - Customer Type = "business"
          - Customer Size = "medium"
            - location = "center": no (40/0)
    - sales person = "c"
      - Customer Type = "business"
        - Customer Size = "large"
          - Contact-Initiation = "Company"
            - location = "close-periphery": no (40/0)
  - reference = "yes"
    - sales person = "a": yes (40/0)
    - sales person = "b"
      - Customer Type = "business"
        - Customer Size = "medium": yes (200/20)
    - sales person = "c"
      - cost-offer > 300
        - Contact-Initiation = "Company"
          - Customer Type = "business": yes (120/40)
            - Contact-Initiation = "Customer": yes (40/0)
      - cost-offer <= 300
        - Customer Type = "private"
          - Customer Size = "small": yes (160/20)

Number	of		leaves:	21
Number	of		attributes:	73
Total		Benefit:		32000
Tree		Accuracy:		73.25%
Correctly	Classified		Instances:	1240
Incorrectly Classified Instances: 360				

Fig. 5.5 (continued)

on the domain specific knowledge it receives. This is the novelty of the proactive data mining method.

In this scenario the optimization algorithm looks for the path that maximizes the utility function given the different costs and benefits that are provided (Matatov et al. 2010; Menahem et al. 2009). It is clear that the model consists of more paths that are constructed with changeable attributes located in places which provide considerable

Table 5.7 Optimization algorithm generated action list over the Maximal Utility model for the security company with dynamic benefit function and infinite change cost for the unchangeable attributes

From Branch (β_i)	To Branch (β_j)	utility (β_i, β_j)
Call or Visit = "call" AND Contact-Initiation = "Company" AND cost-offer > 300 AND sales person = "a" AND Customer Size = "medium" AND Customer Type = "business"	Call or Visit = "visit" AND reference = "yes" AND sales person = "a"	195078.26
Call or Visit = "call" AND Contact-Initiation = "Company" AND cost-offer > 300 AND sales person = "a" AND Customer Size = "small" AND Customer Type = "business"	Call or Visit = "visit" AND reference = "yes" AND sales person = "a"	167209.93
Call or Visit = "visit" AND reference = "yes" AND sales person = "c" AND cost-offer ≤ 300 AND Customer Type = "private" AND Customer Size = "small"	Call or Visit = "visit" AND reference = "yes" AND sales person = "a"	115189.07
Call or Visit = "call" AND Contact-Initiation = "Customer" AND sales person = "b" AND reference = "no" AND location = "far-periphery" AND Customer Type = "private"	Call or Visit = "visit" AND reference = "yes" AND sales person = "a"	111473.29
Call or Visit = "call" AND Contact-Initiation = "Customer" AND sales person = "a" AND Customer Type = "private" AND Customer Size = "small" AND location = "center-center"	Call or Visit = "visit" AND reference = "yes" AND sales person = "a"	111473.29
Call or Visit = "visit" AND reference = "yes" AND sales person = "c" AND cost-offer > 300 AND Contact-Initiation = "Company" AND Customer Type = "business"	Call or Visit = "visit" AND reference = "yes" AND sales person = "a"	74315.53
Call or Visit = "visit" AND reference = "yes" AND sales person = "b" AND Customer Type = "business" AND Customer Size = "medium"	Call or Visit = "visit" AND reference = "yes" AND sales person = "a"	74315.53
Call or Visit = "visit" AND reference = "no" AND sales person = "c" AND Customer Type = "business" AND Customer Size = "large" AND Contact-Initiation = "Company" AND location = "close-periphery"	Call or Visit = "visit" AND reference = "yes" AND sales person = "a"	74315.53
Call or Visit = "call" AND Contact-Initiation = "Company" AND cost-offer ≤ 300 AND sales person = "c" AND Customer Type = "private" AND Customer Size = "small"	Call or Visit = "visit" AND reference = "yes" AND sales person = "a"	74315.53
Call or Visit = "visit" AND reference = "no" AND sales person = "b" AND Contact-Initiation = "Customer" AND Customer Type = "business" AND Customer Size = "medium" AND location = "center"	Call or Visit = "visit" AND reference = "yes" AND sales person = "a"	65026.085
Call or Visit = "call" AND Contact-Initiation = "Company" AND cost-offer σσσ 300 AND sales person = "c" AND Customer Type = "business" AND Customer Size = "medium"	Call or Visit = "visit" AND reference = "yes" AND sales person = "a"	65026.08
Call or Visit = "call" AND Contact-Initiation = "Customer" AND sales person = "b" AND reference = "no" AND location = "center-center" AND Customer Type = "private" AND Customer Size = "small"	Call or Visit = "visit" AND reference = "yes" AND sales person = "a"	48305.09

Table 5.7 (continued)

From Branch (β_i)	To Branch (β_j)	utility (β_i, β_j)
Call or Visit = "call" AND Contact-Initiation = "Company" AND cost-offer ≤ 300 AND sales person = "b" AND Customer Type = "private" AND Customer Size = "small" AND location = "center-center"	Call or Visit = "visit" AND reference = "a" AND sales person = "a"	48305.09
Call or Visit = "visit" AND reference = "no" AND sales person = "b" AND Contact-Initiation = "Company" AND Customer Type = "business"	Call or Visit = "visit" AND reference = "a" AND sales person = "a"	37157.76
Call or Visit = "visit" AND reference = "no" AND sales person = "a" AND Customer Type = "business" AND Customer Size = "medium" AND Contact-Initiation = "Customer"	Call or Visit = "visit" AND reference = "a" AND sales person = "a"	37157.76
Call or Visit = "visit" AND reference = "no" AND sales person = "a" AND Customer Type = "business" AND Customer Size = "large" AND Contact-Initiation = "Company"	Call or Visit = "visit" AND reference = "a" AND sales person = "a"	37157.76
Call or Visit = "call" AND Contact-Initiation = "Customer" AND sales person = "b" AND reference = "yes" AND Customer Type = "private" AND Customer Size = "small"	Call or Visit = "visit" AND reference = "a" AND sales person = "a"	37157.76
Call or Visit = "call" AND Contact-Initiation = "Company" AND cost-offer ≤ 300 AND Customer Type = "private" AND Customer Size = "small" person = "a" AND Customer Type = "private" AND Customer Size = "small"	Call or Visit = "visit" AND reference = "a" AND sales person = "a"	37157.76
Call or Visit = "call" AND Contact-Initiation = "Company" AND cost-offer > 300 AND sales person = "b" AND Customer Type = "business" AND Customer Size = "small"	Call or Visit = "visit" AND reference = "a" AND sales person = "a"	37157.76
Call or Visit = "visit" AND reference = "yes" AND sales person = "c" AND cost-offer > 300 AND Contact-Initiation = "Customer"	Call or Visit = "visit" AND reference = "a" AND sales person = "a"	9289.44

Table 5.8 Generated action list over the J48 model for the security company with dynamic benefit function and infinite change cost for the unchangeable attributes

From Branch (β_i)	To Branch (β_j)	utility (β_i, β_j)
Call or Visit = "call"	Call or Visit = "visit" AND reference = "yes"	482840.74
Call or Visit = "visit" AND reference = "no" AND sales person = "c"	Call or Visit = "visit" AND reference = "yes"	48780.14
Call or Visit = "visit" AND reference = "no" AND sales person = "b" AND Customer Size = "medium"	Call or Visit = "visit" AND reference = "yes"	40512.32
Call or Visit = "visit" AND reference = "no" AND sales person = "a"	Call or Visit = "visit" AND reference = "yes"	31417.72

Table 5.9 Cost matrix for contact-initiation

	Customer	Company
Customer	0	0
Company	250	0

Table 5.10 Cost matrix for call-or-visit

	Call	Visit
Call	0	375
Visit	0	0

Table 5.11 Cost matrix for reference

	Yes	No
Yes	0	0
No	90	0

Table 5.12 Cost matrix for salesperson

	A	b	c
a	0	325	230
b	475	0	200
c	500	300	0

room for potential beneficial changes. For example, the sub-trees under each of the salesperson attributes (i.e., a changeable attribute) allow for potential actions. It is important to note, that even when a source path includes unchangeable attribute(s) it can be moved to a target path that does not include the unchangeable attribute(s). However, all records classified under the target path must have the same unchangeable attribute value as in the source path. Table 5.13 lists all suggested potential actions by the optimization algorithm running on the Maximal Utility model presented in Fig. 5.6.

Although the target path [reference = "yes" AND Call or Visit = "visit" AND sales person = "a"] is the preferred target path, it is not the single dominant path. Since this preferred path consists of changeable attributes only and has a 100 % success rate (i.e., 40/0 yes), the optimization algorithm tries to "move" to it as many source paths as possible. However, given the balanced costs, it is not the most beneficial target path for all source paths. The total potential utility gain from the Maximal

- reference = "no"
  - Customer Type = "business"
    - sales person = "a"
      - Customer Size = "large": no (40/20)
      - Customer Size = "medium"
        - Contact-Initiation = "Company": no (120/0)
        - Contact-Initiation = "Customer": no (40/20)
      - Customer Size = "small": no (120/0)
    - sales person = "b"
      - Contact-Initiation = "Company": yes (80/20)
      - Contact-Initiation = "Customer"
        - Customer Size = "medium": no (40/0)
    - sales person = "c"
      - Customer Size = "large"
        - Contact-Initiation = "Company": no (40/0)
  - Customer Type = "private"
    - location = "center-center": no (160/60)
    - location = "far-periphery": no (120/60)
- reference = "yes"
  - Call or Visit = "call"
    - Customer Size = "medium"
      - Customer Type = "business"
        - Contact-Initiation = "Company": no (40/0)
    - Customer Size = "small"
      - Customer Type = "business"
        - Contact-Initiation = "Company": no (40/20)
      - Customer Type = "private": no (200/80)
  - Call or Visit = "visit"
    - sales person = "a": yes (40/0)
    - sales person = "b": yes (200/20)
    - sales person = "c"
      - Customer Size = "large": yes (120/40)
      - Customer Size = "medium": yes (40/0)
      - Customer Size = "small": yes (160/20)

Number of leaves: 17
Number of attributes: 34
Total Benefit: 32000
Tree Accuracy: 75.37%
Correctly Classified Instances: 1240
Incorrectly Classified Instances: 360

Fig. 5.6 Maximal Utility model for the security company with infinite change cost for the unchangeable attributes, balanced costs for the changeable attributes and dynamic benefit function

Utility model with the dynamic benefit function, an infinite change cost for the unchangeable attributes, and balanced costs for the changeable attributes is: utility $(\beta_i, \beta_j) = 494{,}576.76$. This potential utility gain is lower than the one achieved in

Table 5.13 Optimization algorithm action list over the Maximal Utility model for the security company with dynamic benefit function, infinite change cost for the unchangeable attributes and balanced change cost for the changeable attributes

From Branch (β_i)	To Branch (β_j)	utility (β_i, β_j)
Reference = "no" AND Customer Type = "business" AND sales person = "a" AND Customer Size = "medium" AND Contact-Initiation = "Company"	Reference = "yes" AND Call or Visit = "visit" AND sales person = "a"	111078.26
Reference = "no" AND Customer Type = "business" AND sales person = "a" AND Customer Size = "small"	Reference = "yes" AND Call or Visit = "visit" AND sales person = "a"	83209.93
Reference = "yes" AND Call or Visit = "call" AND Customer Size = "small" AND Customer Type = "private"	Reference = "yes" AND Call or Visit = "visit" AND sales person = "a"	43936.14
Reference = "no" AND Customer Type = "business" AND sales person = "c" AND Customer Size = "large" AND Contact-Initiation = "Company"	Reference = "yes" AND Call or Visit = "visit" AND sales person = "a"	41315.53
Reference = "yes" AND Call or Visit = "call" AND Customer Size = "medium" AND Customer Type = "business" AND Contact-Initiation = "Company"	Reference = "yes" AND Call or Visit = "visit" AND sales person = "c" AND Customer Size = "medium"	40736.64
Reference = "yes" AND Call or Visit = "visit" AND sales person = "c" AND Customer Size = "small"	Reference = "yes" AND Call or Visit = "visit" AND sales person = "a"	35189.066
Reference = "no" AND Customer Type = "business" AND sales person = "b" AND Contact-Initiation = "Customer" AND Customer Size = "medium"	Reference = "yes" AND Call or Visit = "visit" AND sales person = "a"	34736.64
Reference = "no" AND Customer Type = "private" AND location = "center-center"	Reference = "yes" AND Call or Visit = "visit" AND sales person = "a"	28778.38
Reference = "no" AND Customer Type = "business" AND sales person = "a" AND Customer Size = "medium" AND Contact-Initiation = "Customer"	Reference = "yes" AND Call or Visit = "visit" AND sales person = "a"	24157.76
Reference = "no" AND Customer Type = "business" AND sales person = "a" AND Customer Size = "large"	Reference = "yes" AND Call or Visit = "visit" AND sales person = "a"	24157.76
Reference = "yes" AND Call or Visit = "visit" AND sales person = "c" AND Customer Size = "large"	Reference = "yes" AND Call or Visit = "visit" AND sales person = "a"	14315.53
Reference = "yes" AND Call or Visit = "call" AND Customer Size = "small" AND Customer Type = "business" AND Contact-Initiation = "Company"	Reference = "no" AND Customer Type = "business" AND sales person = "b" AND Contact-Initiation = "Company"	12965.11

Table 5.14 Optimization algorithm action list over the J48 model for the security company with dynamic benefit function, infinite change cost for the unchangeable attributes and balanced change cost for the changeable attributes

From Branch (β_i)	To Branch (β_j)	utility (β_i, β_j)
Call or Visit = "visit" AND reference = "no" AND sales person = "c"	Call or Visit = "visit" AND reference = "yes"	35780.14
Call or Visit = "visit" AND reference = "no" AND sales person = "b" AND Customer Size = "medium"	Call or Visit = "visit" AND reference = "yes"	27512.32
Call or Visit = "call"	Call or Visit = "visit" AND reference = "yes"	13840.74
Call or Visit = "visit" AND reference = "no" AND sales person = "a"	Call or Visit = "visit" AND reference = "yes"	5417.72

the previous scenario (Table 5.7) due to the change cost that has been assigned to the changeable attributes. On the other hand, running the optimization algorithm with the dynamic benefit function, infinite change cost for the unchangeable attributes, and balanced costs for the changeable attributes over the J48 model (Fig. 5.3) will produce the suggested action list presented in Table 5.14.

As we can see from the above list, the optimization algorithm finds the possible actions that include only changeable attributes. The total potential utility gain from the J48 model with the dynamic benefit function, infinite change cost for the unchangeable attributes, and balanced costs for the changeable attributes is: utility (β_i, β_j) = 82,550. This potential utility gain is much lower than the one achieved in the previous scenario (Table 5.8) due to the costs for the changeable attributes constraint.

5.5   Chapter Summary

From the above scenarios, we can state that when there are no limits or constraints (i.e., no cost), the Maximal Utility generated models are larger to allow for as many optimization actions as possible. The more constraints, the smaller the decision model and the lower the potential total utilities gain. Accordingly, the Maximal Utility generated model in Fig. 5.4 consists of 83 attributes and a potential total utility gain of 1,456,584.31; Fig. 5.5 has 73 attributes and a potential utility gain 1,456,584.31; and Fig. 5.6 has only 34 attributes and a potential total utility gain of 494,576.

References

Ben-Shimon D, Tsikinovsky A, Rokach L, Meisles A, Shani G, Naamani L (2007) Recommender system from personal social networks. In: Wegrzyn-Wolska K , Szczepaniak P. (eds), Advances in Intelligent Web Mastering. Advances in Soft Computing, vol 43, Springer Berlin Heidelberg, pp 47–55

Breiman L (1996) Bagging predictors. Mach Learn 24:123–140

Buntine W, Niblett T (1992) A further comparison of splitting rules for decision-tree induction. Mach Learn 8:75–85

Fayyad U, Irani KB (1992) The attribute selection problem in decision tree generation. In: Proceedings of tenth national conference on artificial intelligence. AAAI Press/MIT Press, Cambridge, pp 104–110

Kisilevich S, Rokach L, Elovici Y, Shapira B (2010) Efficient multidimensional suppression for k-anonymity. IEEE Trans Knowledge Data Eng 22(3):334–347

Lim T-S, Loh W-Y, Shih Y-S (2000) A comparison of prediction accuracy, complexity, and training time of thirty-three old and new classification algorithms. Mach Learn 40(3):203–228

Loh WY, Shih X (1997) Split selection methods for classification trees. Statistica Sinica 7:815–840

Loh WY, Shih X (1999) Families of splitting criteria for classification trees. Stat Comput 9:309–315

Maimon O, Rokach L (2001) Data mining by attribute decomposition with semiconductor manufacturing case study. In: Braha D (ed) Data mining for design and manufacturing. Massive Computing, vol 3. Springer US, pp 311–336

Matatov N, Rokach L, Maimon O (2010) Privacy-preserving data mining: a feature set partitioning approach. Info Sci 180(14):2696–2720

Menahem E, Rokach L, Elovici Y (2009) Troika–An improved stacking schema for classification tasks. Info Sci 179(24):4097–4122

Provost F, Fawcett T (1997) Analysis and visualization of Classifier Performance comparison under imprecise class and cost distribution. In: Proceedings of KDD-97. AAAI Press, pp 43–48

Provost F, Fawcett T (1998) The case against accuracy estimation for comparing induction algorithms. In: Proc. 15th Intl. Conf. on machine learning, Madison, WI, pp 445–453

Rokach L (2006) Decomposition methodology for classification tasks: a meta decomposer framework. Pattern Anal Appl 9(2–3):257–271

Rokach L (2009) Collective-agreement-based pruning of ensembles. Comput Stat Data Anal 53(4):1015–1026

Chapter 6
Conclusions

Most works on data mining focus on methods for extracting patterns from datasets containing data from the past or previous events. Although useful, the ability to extract patterns by itself does not provide a holistic answer to what businesses really need—optimization rather than merely discovery.

There is a gap between academic literature and business applications this springs from three shortcomings of traditional data mining methods. (i) Most traditional data mining methods are passive rather than proactive; (ii) they ignore relevant environmental knowledge, and (iii) they are focused on model accuracy. There are very few, if any, data mining methods that overcome all these shortcomings altogether.

To overcome these shortcomings, this book proposes *Proactive Data Mining with Decision Trees* a novel, proactive approach to data-mining. In particular, this book suggests a specific implementation of the novel domain-driven proactive approach for classification trees. The new approach not only induces a model for predicting or explaining a phenomenon, but also utilizes specific problems/domain knowledge to suggest specific actions for achieving optimal changes in the value of the target attribute.

Domain-driven proactive classification is a two phase process. In the first phase it trains a probabilistic classifier using a supervised learning algorithm. The resulting classification model from the first phase is a model that is predisposed to potential interventions and is oriented toward maximizing various utility functions the organization may set. In the second phase it utilizes the induced classifier to suggest potential actions for maximizing utility while reducing costs.

Unlike previous post-processing methods that use existing classification algorithms, the methods presented in this book consider the pursuit of utility-maximization as an integral part of the core learning algorithm. It is clearly proven that data mining algorithms that inherently consider utility-maximization are potentially much more profitable than models that post-process the results that traditional data mining algorithms provide.

In this book we demonstrate that by taking the domain-driven proactive classification approach, it becomes possible to solve business problems which cannot be approach by traditional, passive data-mining methods. More specifically, when considering the distribution of the input observations as changeable, business users can

H. Dahan et al., *Proactive Data Mining with Decision Trees,*
SpringerBriefs in Electrical and Computer Engineering,
DOI 10.1007/978-1-4939-0539-3_6, © The Author(s) 2014

evaluate the outcomes of actions that change this distribution. The domain knowledge provides the basis for measuring these outcomes. The case studies presented in Chap. 4, demonstrate how this approach can provide significant competitive value in stormy business environments.

In addition to the two-phase framework, the book also suggests a novel splitting criteria, for decision trees. We show that the proposed splitting criterion is superior in that it tends to produce decision trees with higher potential for utility enhancement. We demonstrated that a narrow focus on classification accuracy has little to do with the business objective of optimization.

We conclude this book with several suggestions for future research:

(i) Case studies: The approach that was proposed in this work was triggered by observing the need of many businesses not for theoretical knowledge but for practical tools that could be used to optimize and enhance business activities. We believe that in order to better understand the proposed approach and to reduce the gap between the theories that academia proposes and the applications that businesses require, we need to implement the proposed algorithms on additional real-life case studies.

(ii) Since the algorithms presented in this book are based on domain-knowledge, and not merely on a training dataset, it is difficult to compare their performance with existing methods. Accordingly, we suggest constructing a series of measures to evaluate the results of the proactive domain-driven classification algorithms (similar to the way that zero-one loss is used to evaluate traditional classification methods).

(iii) In this book we focused applying the proposed approach to decision trees. Decision trees have several advantages in this respect. However, we suggest developing the domain-driven proactive classification approach to support data-mining algorithms other than decision-trees (i.e., neural networks).

(iv) In this book we consider a specific form of domain knowledge (benefits and action-change costs). Since various businesses may need other forms of domain knowledge, we suggest implementing the approach to wider forms of domain knowledge.

(v) During our research, we found the algorithms proposed in this book help in solving other data mining problems (i.e., feature selection, cost-proportionate weighting of the training example, etc.). We suggest exploring these opportunities.